Computational Methods of
Multi-Physics Problems

Computational Methods of Multi-Physics Problems

Special Issue Editor

Timon Rabczuk

MDPI • Basel • Beijing • Wuhan • Barcelona • Belgrade

MDPI

Special Issue Editor
Timon Rabczuk
Bauhaus University Weimar
Germany

Editorial Office
MDPI
St. Alban-Anlage 66
4052 Basel, Switzerland

This is a reprint of articles from the Special Issue published online in the open access journal *Energies* (ISSN 1996-1073) from 2018 to 2019 (available at: https://www.mdpi.com/journal/energies/special_issues/multiphysics)

For citation purposes, cite each article independently as indicated on the article page online and as indicated below:

LastName, A.A.; LastName, B.B.; LastName, C.C. Article Title. *Journal Name* **Year**, *Article Number*, Page Range.

ISBN 978-3-03921-417-4 (Pbk)
ISBN 978-3-03921-418-1 (PDF)

Contents

About the Special Issue Editor vii

Preface to "Computational Methods of Multi-Physics Problems" ix

Hanlin Wang, Erkan Oterkus and Selda Oterkus
Three-Dimensional Peridynamic Model for Predicting Fracture Evolution during the
Lithiation Process
Reprinted from: *Energies* **2018**, *11*, 1461, doi:10.3390/en11061461 1

**Timon Rabczuk, Mohammad Reza Azadi Kakavand, Raahul Palanivel Uma,
Ali Hossein Nezhad Shirazi and Meysam Makaremi**
Thermal Conductance along Hexagonal Boron Nitride and Graphene Grain Boundaries
Reprinted from: *Energies* **2018**, *11*, 1553, doi:10.3390/en11061553 23

Bohayra Mortazavi and Timon Rabczuk
Boron Monochalcogenides; Stable and Strong Two-Dimensional Wide
Band-Gap Semiconductors
Reprinted from: *Energies* **2018**, *11*, 1573, doi:10.3390/en11061573 37

Reza Khademi-Zahedi and Pouyan Alimouri
Finite Element Analysis to the Effect of Thermo-Mechanical Loads on Stress Distribution in
Buried Polyethylene Gas Pipes Jointed by Electrofusion Sockets, Repaired by PE Patches
Reprinted from: *Energies* **2018**, *11*, 2818, doi:10.3390/en11102818 47

Bo He, Brahmanandam Javvaji and Xiaoying Zhuang
Characterizing Flexoelectricity in Composite Material Using the Element-Free Galerkin Method
Reprinted from: *Energies* **2019**, *12*, 271, doi:10.3390/en12020271 71

José Reinoso, Percy Durand, Pattabhi Ramaiah Budarapu and Marco Paggi
Crack Patterns in Heterogenous Rocks Using a Combined Phase Field-Cohesive Interface
Modeling Approach: A Numerical Study
Reprinted from: *Energies* **2019**, *12*, 965, doi:10.3390/en12060965 89

About the Special Issue Editor

Timon Rabczuk obtained his PhD from the University of Karlsruhe. He worked at the Fraunhofer Institute (Ernst Mach) in Freiburg before joining the Computational Mechanics Group of Prof. Ted Belytschko at Northwestern University in Evanston, USA, where he worked for 4 years as Postdoctoral Fellow. For 1.5 years, Prof. Rabczuk was a member of the Computational Mechanics group at the Technical University of Munich. In February 2007, he was appointed to the position of Senior Lecturer at the Department of Mechanical Engineering at Canterbury University, Christchurch, New Zealand. In 2009, Prof. Rabczuk joined the Bauhaus University Weimar as Full Professor.

Preface to "Computational Methods of Multi-Physics Problems"

Multi-physics problems play a huge role in engineering. They include hydraulic fracturing, piezoelectricity, flexoelectricity, energy harvesters, energy storage systems and batteries, just to name a few relevant applications. In order to support the design of materials, systems and devices, computational modeling is of major importance. It can complement experimental testing and shed light on physical phenomena that are difficult to detect or measure experimentally. Furthermore, computational modeling can be used to—more systematically—virtually design a material or a component, saving cost and time. This Special Issue aims to reveal the state-of-the-art in addressing computational models for the computational modeling of multi-field problems.

Timon Rabczuk
Special Issue Editor

energies

MDPI

Article

Three-Dimensional Peridynamic Model for Predicting Fracture Evolution during the Lithiation Process

Hanlin Wang, Erkan Oterkus * and Selda Oterkus

Department of Naval Architecture, Ocean and Marine Engineering, University of Strathclyde, Glasgow G4 0LZ, UK; hanlin.wang.2013@uni.strath.ac.uk (H.W.); selda.oterkus@strath.ac.uk (S.O.)
* Correspondence: erkan.oterkus@strath.ac.uk; Tel.: +44-141-548-3876

Received: 12 May 2018; Accepted: 31 May 2018; Published: 5 June 2018

Abstract: Due to its large electric capacity, silicon has become one of the most promising electrode materials for lithium ion batteries. However, silicon undergoes large volumetric expansion and material stiffness reduction during the charging process. This can lead to fracture and failure of lithium-ion batteries. Damage formation and evolution inside the electrode are influenced by the lithium ion concentration and electrode material. High stress gradients induced by heterogeneous deformation can lead to massive migration of lithium ions towards high geometrical singularity regions, such as crack edge regions, which increases the lithium ion concentration. Fully coupled mechanical diffusion equations are important in describing the mechanics of this problem. In this study, the three-dimensional peridynamic theory is presented to solve the coupled field problem. In addition, the newly developed peridynamic differential operator concept is utilized to convert partial differential equations into peridynamic form for the diffusion equation. Spherical and cylindrical shaped energy storage structures with different pre-existing penny-shaped cracks are considered to demonstrate the capability of the developed framework. It is shown that peridynamic theory is a suitable tool for predicting crack evolution during the lithiation process.

Keywords: lithium-ion battery; fracture analysis; peridynamics; pressure gradient effect

1. Introduction

In the marine industry, traditional propulsion systems with internal combustion engines cannot match the increasing strict standards for waste gas emissions (MARPOL Annex VI) [1]. Hence, the hybrid propulsion system, as an alternative approach to marine propulsion systems, has become popular in the shipping industry. Marine batteries, as primary energy storage devices, have been subject to excessive investigation [2]. The lithium ion battery is one of the most promising marine batteries due to its high energy to weight ratio, high energy density, low rate of self-discharge and long cycle life [3]. The performance of a lithium ion battery mainly depends on the material properties of its anode, cathode and electrolyte. The carbon group elements, carbon (C), silicon (Si), germanium (Ge), tin (Sn) and their alloys, can be selected as the anode material in lithium ion batteries [4,5]. In many commercial grade lithium ion batteries, graphitic carbon is the main component of the anode material due to its low expansion induced by lithiation during battery charging [4]. However, the major limitation of graphite carbon is its low electric capacity (372 mAhg^{-1}) whereby it stores the lithium ion inside graphite sheets as (LiC$_6$) [4–6] which cannot satisfy the demand for high electric capacity in marine batteries. On the other hand, silicon (Si) and germanium (Ge) have relatively high theoretical capacities as compared to graphite carbon (3579 mAhg^{-1} and 1625 mAhg^{-1}, respectively) [6] which can be considered to be suitable candidates for the anode material in lithium ion batteries. However, both Si and Ge anodes experience large volumetric changes during battery cycling. The Si anode can experience a volume expansion of up to 400% during the charging process [7,8]. Similar to the Si anode, the Ge anode can expand 370% in volume during the lithiation process [9]. Frequent cycling of the lithium ion battery may lead to stress misdistribution, degradation and delamination of the battery components. This may then lead to failure or even pulverization of the battery anode.

Many efforts have been conducted to increase the electric capacity and to avoid the occurrence of defects inside the electrodes of lithium ion batteries. Winter et al. [10] suggested that some metals, such as aluminum (Al), tin (Sn) and antimony (Sb) can store more lithium ion than graphite carbon by forming alloys with lithium ions. However, the newly formed alloyed anode will experience a 500% volumetric expansion [4], which is the major barrier to becoming the anode material in rechargeable lithium ion batteries. Gao et al. [11] suggested carbon nanotubes as the anode material. Carbon nanotubes are tubular forms of graphite sheet with high conductivity, high tensile strength, high rigidity and low density. They increase the electric capacity of carbon by up to around $600\,\text{mAhg}^{-1}$ without damage or pulverization. Metals or their alloys can also be added into the nanotube in composite form. Hence, the advantage of the high electric capacity of metal alloys and the low volumetric change in graphite carbon can be utilized in anode materials [12,13]. However, carbon nanotubes experience a loss of lithium ion capacity during cycling and a linear voltage drop during discharging [4].

In comparison with graphite carbon or carbon nanotube anodes, silicon has superior performance in terms of electrical capacity. However, as mentioned earlier, it suffers from pulverization due to large volumetric change during battery cycling. The fracture behaviour of silicon anodes has been subject to frequent investigation in recent years. Liu et al. [14] constructed a thin silicon film model to investigate lithiation-induced tensile stress and surface cracking by using analytical and finite element methods (FEM). A compression-traction transition zone was observed along the interface of electrode and electrolyte. The formation of the transition zone depends on the large volumetric change, plastic deformation and slow charging rate. Crack formation and propagation was led by the location of the transition zone [14,15]. The magnitude and profiles of tensile stress at the surface of the lithiated silicon zone depends on volumetric misfit strain, yield stress and modulus of unlithiated silicon. Ryu et al. [8] found that the large volume change during the normal cycling process is always accompanied by a pressure gradient. High hydrostatic stress will affect the lithium ion diffusion and fracture propagation inside the anode of a battery. Grantab and Shenoy [16] conducted a detailed investigation of the effect of the pressure gradient factor on crack propagation in silicon nanowires. The cohesive FEM was applied to model fracture evolution in silicon nanowires. Since localized stress around the crack tip is higher than the surrounding nanowire surface region, a large quantity of lithium ions moved into the crack tip region which caused a relatively large volume expansion. Therefore, the hydrostatic stress around the crack tip reduced. Zuo and Zhao [17] used the phase field method to study the stress evolution and crack propagation. Several damaged anode models with different crack numbers and different crack orientations were considered to illustrate fracture evolution in the battery anode. The pressure gradient was shown to be dependent on the elastic modulus, partial molar volume, concentration, Poisson's ratio and the localized lithium ion concentration. On the other hand, Gao and Zhou [18] applied the FEM and the J-integral method to study the softening effect during the lithiation process. They also captured the high lithium ion concentration at crack tip regions during the charging process, leading to relaxation of hydrostatic stress in later diffusion process. The fracture evolution in silicon nanowires also depends on the geometry of the structure. Ryu et al. [8] also concluded that silicon nanowires with diameters smaller than 300 nm will not fail during battery cycling, even pre-damaged nanowires. Due to their large pressure gradient and large volume expansion, lithium ions can rapidly diffuse into silicon nanowires. As an alternative approach, peridynamics [19–25] was used to investigate fracture evolution during the lithiation process by considering plane-stress conditions [26,27]. It was shown that peridynamics can capture accurate fracture propagation patterns for multiple cracks with random orientations inside the anode material.

In this study, the two-dimensional fully-coupled peridynamic model developed in Wang et al. [27] is extended to a three-dimensional peridynamic model. The effect of pressure gradient and material phase change during lithiation process (from Si to Li$_x$Si) are taken into consideration. Coupled field equations are used to describe the lithiation process with the help of the newly developed peridynamic differential operator [26]. To demonstrate the capability of the model, the fracture evolution inside the cylindrical and spherical shaped lithium ion battery anodes is presented.

2. Coupled Diffusion-Mechanical Deformation Formulation

Fick's Second Law describes the basic principles of ion diffusion [28]. Due to silicon being the anode material, larger volume expansion occurs as lithium ions diffuse inside the anode. Hence, the diffusion-induced stress during battery cycling should be taken into account in Fick's Second Law. The mechanical strain component, ε_{ij}^M, can be expressed by considering the influence of lithium ion diffusion as [16,17]:

$$\varepsilon_{ij}^M = \varepsilon_{ij} - \alpha C_{avg}\delta_{ij} \tag{1}$$

where ε_{ij} refers to the total strain, α represents the coefficient of expansion, C_{avg} represents the current average concentration value of the interacting material points, and δ_{ij} is the Kronecker Delta. For three-dimensional cases, the stress-strain constitutive equation can be written as:

$$\begin{Bmatrix} \sigma_{xx} \\ \sigma_{yy} \\ \sigma_{zz} \\ \sigma_{xy} \\ \sigma_{xz} \\ \sigma_{yz} \end{Bmatrix} = \frac{E}{(1+\nu)(1-2\nu)} \begin{bmatrix} 1-\nu & \nu & \nu & 0 & 0 & 0 \\ \nu & 1-\nu & \nu & 0 & 0 & 0 \\ \nu & \nu & 1-\nu & 0 & 0 & 0 \\ 0 & 0 & 0 & \frac{1-2\nu}{2} & 0 & 0 \\ 0 & 0 & 0 & 0 & \frac{1-2\nu}{2} & 0 \\ 0 & 0 & 0 & 0 & 0 & \frac{1-2\nu}{2} \end{bmatrix} \begin{Bmatrix} \varepsilon_{xx}^M \\ \varepsilon_{yy}^M \\ \varepsilon_{zz}^M \\ 2\varepsilon_{xy}^M \\ 2\varepsilon_{xz}^M \\ 2\varepsilon_{yz}^M \end{Bmatrix} \tag{2}$$

By substituting Equation (2) into Equation (1), the coupled field normal stresses can be expressed as:

$$\sigma_{xx} = \frac{E}{(1+\nu)(1-2\nu)}\left[(1-\nu)\varepsilon_{xx} + \nu\varepsilon_{yy} + \nu\varepsilon_{zz}\right] - \frac{E}{(1-2\nu)}\alpha C_{max}C \tag{3}$$

$$\sigma_{yy} = \frac{E}{(1+\nu)(1-2\nu)}\left[\nu\varepsilon_{xx} + (1-\nu)\varepsilon_{yy} + \nu\varepsilon_{zz}\right] - \frac{E}{(1-2\nu)}\alpha C_{max}C \tag{4}$$

$$\sigma_{zz} = \frac{E}{(1+\nu)(1-2\nu)}\left[\nu\varepsilon_{xx} + \nu\varepsilon_{yy} + (1-\nu)\varepsilon_{zz}\right] - \frac{E}{(1-2\nu)}\alpha C_{max}C \tag{5}$$

where C is current value of the normalized lithium ion concentration, and C_{max} is the maximum value of the lithium ion concentration. The total normal strain in all three dimensions can be expressed in terms of displacements as:

$$\varepsilon_{xx} = \frac{\partial u}{\partial x} \tag{6}$$

$$\varepsilon_{yy} = \frac{\partial v}{\partial y} \tag{7}$$

$$\varepsilon_{zz} = \frac{\partial w}{\partial z} \tag{8}$$

Local stress will rise along with lithium ion diffusion at regions with high geometrical singularity. A high pressure gradient will lead to a large amount of lithium ion diffusion into these regions, which will increase the lithium ion concentration. As a result, stress will release as volume expands. Hence, the general Fick's Second Law should be modified by considering the pressure-gradient as:

$$
\begin{aligned}
\frac{\partial C}{\partial t} = Mk_B T\left(\frac{\partial^2 C}{\partial x^2} + \frac{\partial^2 C}{\partial y^2} + \frac{\partial^2 C}{\partial z^2}\right) - \frac{MC\Omega}{N_A}\left(\frac{\partial^2 \tilde{\sigma}}{\partial x^2} + \frac{\partial^2 \tilde{\sigma}}{\partial y^2} + \frac{\partial^2 \tilde{\sigma}}{\partial z^2}\right) \\
- \frac{MC\Omega}{N_A}\left(\frac{\partial C}{\partial x}\frac{\partial \tilde{\sigma}}{\partial x} + \frac{\partial C}{\partial y}\frac{\partial \tilde{\sigma}}{\partial y} + \frac{\partial C}{\partial z}\frac{\partial \tilde{\sigma}}{\partial z}\right)
\end{aligned}
\tag{9}
$$

where M is the molecular mobility, k_B is Boltzmann constant, T is the absolute temperature, N_A is Avogadro's constant and $\tilde{\sigma}$ is the hydrostatic stress.

Zhang et al. [29] emphasized that silicon has different lithiated stages during the charging process. In the early stage, since lithium ion concentration is low, silicon will transform into partial lithiated silicon ($Li_x Si$). As the lithium ion concentration increases, the partially lithiated silicon will further transform into fully lithiated silicon ($Li_{15}Si_4$). Moreover, since some material properties of lithiated silicon, such as the elastic modulus and fracture toughness, are lower than those of pure silicon, the lithiation process can also be regarded as a material softening process [30]. However, the size of the partially lithiated silicon region in a battery anode is very small compared to the anode's geometry [29]. Therefore, in this study, only fully lithiated silicon was considered.

3. Peridynamic Theory

An extensive number of studies about fracture mechanics based on classic continuum mechanics (CCM) can be found in the literature. The CCM applies the spatial partial differential equations to describe the motion of a material point. Within the framework of CCM, various tools, such as the FEM, boundary element method (BEM) and cohesive zone method (CZM) [31–33], have been applied in numerical fracture analysis. However, since partial differential equations require continuous material geometry, CCM, as a local theory, may have difficulty performing the numerical analysis of problems with discontinuous geometry, such as cracks and kinks. By regarding damage as newly formed boundaries, a remeshing process may be inevitable in numerical analysis [34].

Peridynamic theory (PD) was firstly introduced by Silling and Askari [35] as an alternative numerical approach to CCM. Unlike FEM, the PD uses spatial integral equations instead of partial differential equations to describe the motion of material points. On the other hand, PD is a non-local theory. Material points can build up interactions, called peridynamic bonds, with surrounding neighbors within a certain distance (δ), as shown in Figure 1. The collection of neighboring points is called the horizon (H_x). For neighboring points of material point **x** beyond the size of the horizon, it is assumed that the interactions are too weak and have little influence on point **x**, which can be ignored in a numerical simulation. By integrating the forces between material point **x** and its neighboring points, **x'**, the equation of motion of material point **x** can be calculated and expressed as

$$
\rho\ddot{\mathbf{u}}(\mathbf{x}, t) = \int_{H_x} \mathbf{f}\big(\mathbf{u}(\mathbf{x}', t) - \mathbf{u}(\mathbf{x}, t), \mathbf{x}' - \mathbf{x}\big) dV_{\mathbf{x}'} + \mathbf{b}(\mathbf{x}, t),
\tag{10}
$$

where **b** refers to the body force density and **f** is the pairwise force of each interaction (bond). Note that, in this study, the bond-based peridynamic theory is used for simplicity. However, extension of the formulation to the ordinary-state based formulation is straightforward [34,36].

In bond-based peridynamic theory, the bond force depends on the relative position and relative displacement of the material points associated with the bond which can be calculated as [37]:

$$
\mathbf{f} = \frac{\boldsymbol{\xi} + \boldsymbol{\eta}}{|\boldsymbol{\xi} + \boldsymbol{\eta}|} c\left(s - \alpha C_{avg}\right)\mu(t, \boldsymbol{\xi})
\tag{11}
$$

where the relative position of material points **x** and **x'** can be defined in the undeformed configuration as:

$$
\boldsymbol{\xi} = \mathbf{x}' - \mathbf{x}
\tag{12}
$$

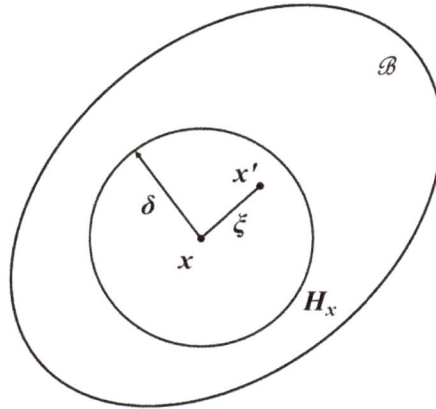

Figure 1. Horizon of material point **x**.

Note that the bond force given in Equation (11) is calculated based on the deformed configuration. Therefore, the formulation has the capacity to capture large deformations. Moreover, the relative displacement can be expressed as:

$$\boldsymbol{\eta} = \mathbf{u}(\mathbf{x}', t) - \mathbf{u}(\mathbf{x}, t) \tag{13}$$

In Equation (11), α is the coefficient of expansion, s is the stretch of the bond and C_{avg} represents the average lithium ion concentration of material points **x** and **x'**. μ is the failure parameter and c is the bond constant. For a three-dimensional isotropic material, the bond constant can be expressed as:

$$c = \frac{12E}{\pi \delta^4} \tag{14}$$

Note that for the three-dimensional, bond-based peridynamic theory, the Poisson's ratio for a linear elastic isotropic material is equivalent to $\nu = 1/4$.

In CCM, damage is specially treated by using remeshing or a pre-defined crack propagation path. On the other hand, the damage is represented as "bond breakage" in peridynamic theory. For brittle elastic material, once the bond stretch exceeds a critical stretch value, s_c, after deformation, the bond will break and cannot recover by itself. The critical stretch depends on the critical strain energy release rate (G_c). For a three-dimensional model, the rate definition of the critical strain energy release in peridynamics has been derived by Silling and Askari [35] as:

$$G_c = \int_0^\delta \left\{ \int_0^{2\pi} \int_0^\delta \int_0^{\cos^{-1} z/\xi} \left(\frac{1}{2} c \xi s_c^2 \xi^2 \right) \sin \phi \, d\phi \, d\xi \, d\theta \right\} dz = \frac{1}{2} c s_c^2 \left(\frac{\delta^5 \pi}{5} \right) \tag{15}$$

Hence, the critical stretch, s_c, can be expressed as:

$$s_c = \sqrt{\frac{5 G_c}{6 E \delta}} \tag{16}$$

The failure condition of each bond can be represented by a failure parameter as:

$$\mu(t, \boldsymbol{\xi}) = \begin{cases} 1 & s \leq s_c \\ 0 & s > s_c \end{cases} \tag{17}$$

Since the broken bond no longer carries any load, the total load associated with material point **x** will be redistributed to the remaining bonds. This leads to a change in mechanical conditions. The damage value of material point **x** can be defined as:

$$\varphi(\mathbf{x}, t) = 1 - \frac{\int_{H_{\mathbf{x}}} \mu(t, \boldsymbol{\xi}) dV_{\mathbf{x}'}}{\int_{H_{\mathbf{x}}} dV_{\mathbf{x}'}} \tag{18}$$

which represents the percentage of broken bonds associated with material point $\mathbf{x}$. The material point is totally damaged when the damage parameter is equivalent to 1, whereas it is undamaged when the damage parameter is equivalent to 0. Note that breakage of a single bond will not cause the emergence of a crack. Instead, all bonds passing through a crack surface should be broken for a crack to occur. In other words, as the number of broken bonds increases, cracks can emerge and eventually propagate inside the structure. Moreover, there is no need to define the crack path, as in some other existing approaches, such as the cracking particle method [38,39].

During the charging (lithiation) process, the anode will expand in volume, and high hydrostatic stresses will emerge inside the anode, especially at high singularity regions. For three-dimensional problems, the hydrostatic stress of a material point, $\mathbf{x}$, can be calculated as [40]:

$$\tilde{\sigma} = \frac{1}{3} \left(\sigma_{xx} + \sigma_{yy} + \sigma_{zz} \right) \tag{19}$$

The Cauchy stress tensor, σ_{ij}, can be expressed in terms of the first Piola–Kirchoff stress, σ_0, as [41]:

$$\boldsymbol{\sigma} = J\boldsymbol{\sigma_0}\mathbf{F}^T = \begin{bmatrix} \sigma_{xx} & \sigma_{xy} & \sigma_{xz} \\ \sigma_{yx} & \sigma_{yy} & \sigma_{yz} \\ \sigma_{zx} & \sigma_{zy} & \sigma_{zz} \end{bmatrix} \tag{20}$$

where the deformation gradient, $\mathbf{F}$, and its determinant, J, are defined as:

$$\mathbf{F} = \frac{9}{4\pi\delta^4} \int_{H_{\mathbf{x}}} \frac{1}{|\mathbf{x}' - \mathbf{x}|^2} (\mathbf{y}' - \mathbf{y}) \otimes (\mathbf{x}' - \mathbf{x}) dV_{\mathbf{x}'} \tag{21}$$

and:

$$J = \det(\mathbf{F}) \tag{22}$$

In Equation (21), $\mathbf{x}' - \mathbf{x}$ refers to the initial relative distance, and $\mathbf{y}' - \mathbf{y}$ refers to the relative distance after deformation. The stress is not directly expressed by the bond-based peridynamic theory. However, the first Piola–Kirchoff stress can be calculated using peridynamic parameters as:

$$\boldsymbol{\sigma_0} = \int_{H_{\mathbf{x}}} \mathbf{f} \otimes (\mathbf{x}' - \mathbf{x}) dV_{\mathbf{x}'} \tag{23}$$

Hence, the amount of hydrostatic stress depends on the displacement gradient and the lithium ion concentration, whereas the concentration change rate depends on the hydrostatic stress gradient and the lithium ion concentration gradient. Therefore, the coupled field equation is applied, as shown in Equation (9). In this study, the peridynamic form of partial differential equation given in Equation (9) was obtained by using peridynamic differential operator method developed by Madenci et al. [42].

4. Peridynamic Differential Operator Approach

The three-dimensional peridynamic differential operator can be derived by considering the Taylor series. Three-dimensional Taylor Series can be written by disregarding the higher order terms [41]:

$$\begin{aligned} f(\mathbf{x}') = f(\mathbf{x}) &+ \xi_1 \frac{\partial f(\mathbf{x})}{\partial x_1} + \xi_2 \frac{\partial f(\mathbf{x})}{\partial x_2} + \xi_3 \frac{\partial f(\mathbf{x})}{\partial x_3} + \frac{1}{2!} \frac{\partial^2 f(\mathbf{x})}{\partial x_1^2} + \frac{1}{2!} \frac{\partial^2 f(\mathbf{x})}{\partial x_2^2} \\ &+ \frac{1}{2!} \frac{\partial^2 f(\mathbf{x})}{\partial x_3^2} + \xi_1\xi_2 \frac{\partial^2 f(\mathbf{x})}{\partial x_1 \partial x_2} + \xi_1\xi_3 \frac{\partial^2 f(\mathbf{x})}{\partial x_1 \partial x_3} + \xi_2\xi_3 \frac{\partial^2 f(\mathbf{x})}{\partial x_2 \partial x_3} \end{aligned} \tag{24}$$

This is done by moving $f(\mathbf{x})$ to the left-hand-side and multiplying each term in Equation (24) with a PD function, $g_3^{p_1 p_2 p_3}(\boldsymbol{\xi})$, in which ($p_1$, p_2, $p_3 = 0$, 1, 2 and they cannot be equal to zero at the same time), and integrating throughout the horizon yields:

$$\int_{H_x} (f(\mathbf{x}') - f(\mathbf{x})) g_3^{p_1 p_2 p_3}(\boldsymbol{\xi}) dV_{\mathbf{x}'} = \frac{\partial f(\mathbf{x})}{\partial x_1} \int_{H_x} \xi_1 g_3^{p_1 p_2 p_3}(\boldsymbol{\xi}) dV_{\mathbf{x}'} + \frac{\partial f(\mathbf{x})}{\partial x_2} \int_{H_x} \xi_2 g_3^{p_1 p_2 p_3}(\boldsymbol{\xi}) dV_{\mathbf{x}'}$$
$$+ \frac{\partial f(\mathbf{x})}{\partial x_3} \int_{H_x} \xi_3 g_3^{p_1 p_2 p_3}(\boldsymbol{\xi}) dV_{\mathbf{x}'} + \frac{\partial^2 f(\mathbf{x})}{\partial x_1^2} \int_{H_x} \frac{1}{2!} \xi_1^2 g_3^{p_1 p_2 p_3}(\boldsymbol{\xi}) dV_{\mathbf{x}'} + \frac{\partial^2 f(\mathbf{x})}{\partial x_2^2} \int_{H_x} \frac{1}{2!} \xi_2^2 g_3^{p_1 p_2 p_3}(\boldsymbol{\xi}) dV_{\mathbf{x}'}$$
$$+ \frac{\partial^2 f(\mathbf{x})}{\partial x_3^2} \int_{H_x} \frac{1}{2!} \xi_3^2 g_3^{p_1 p_2 p_3}(\boldsymbol{\xi}) dV_{\mathbf{x}'} + \frac{\partial^2 f(\mathbf{x})}{\partial x_1 \partial x_2} \int_{H_x} \xi_1 \xi_2 g_3^{p_1 p_2 p_3}(\boldsymbol{\xi}) dV_{\mathbf{x}'}$$
$$+ \frac{\partial^2 f(\mathbf{x})}{\partial x_1 \partial x_3} \int_{H_x} \xi_1 \xi_3 g_3^{p_1 p_2 p_3}(\boldsymbol{\xi}) dV_{\mathbf{x}'} + \frac{\partial^2 f(\mathbf{x})}{\partial x_2 \partial x_3} \int_{H_x} \xi_2 \xi_3 g_3^{p_1 p_2 p_3}(\boldsymbol{\xi}) dV_{\mathbf{x}'} \tag{25}$$

The orthogonality property of the PD functions is invoked:

$$\frac{1}{n_1 n_2 n_3} \int_{H_x} \xi_1^{n_1} \xi_2^{n_2} \xi_3^{n_3} g_3^{p_1 p_2 p_3}(\boldsymbol{\xi}) dV_{\mathbf{x}'} = \delta_{n_1 p_1} \delta_{n_2 p_2} \delta_{n_3 p_3} \text{ with } (n_1, n_2, p_1, p_2 = 1, 2) \tag{26}$$

and Equation (26) is substituted into Equation (25); then, the partial differential equations can be expressed in peridynamic form as:

$$\begin{Bmatrix} \frac{\partial f(\mathbf{x})}{\partial x_1} \\ \frac{\partial f(\mathbf{x})}{\partial x_2} \\ \frac{\partial f(\mathbf{x})}{\partial x_3} \\ \frac{\partial^2 f(\mathbf{x})}{\partial x_1^2} \\ \frac{\partial^2 f(\mathbf{x})}{\partial x_2^2} \\ \frac{\partial^2 f(\mathbf{x})}{\partial x_3^2} \\ \frac{\partial^2 f(\mathbf{x})}{\partial x_1 \partial x_2} \\ \frac{\partial^2 f(\mathbf{x})}{\partial x_1 \partial x_3} \\ \frac{\partial^2 f(\mathbf{x})}{\partial x_2 \partial x_3} \end{Bmatrix} = \int_{Hx} (f(\mathbf{x}') - f(\mathbf{x})) \begin{Bmatrix} g_3^{100} \\ g_3^{010} \\ g_3^{001} \\ g_3^{200} \\ g_3^{020} \\ g_3^{002} \\ g_3^{110} \\ g_3^{101} \\ g_3^{011} \end{Bmatrix} dV_{\mathbf{x}'} \tag{27}$$

In Equation (27), the peridynamic functions, $g_3^{p_1 p_2 p_3}(\boldsymbol{\xi})$, can be constructed by:

$$g_3^{p_1 p_2 p_3}(\boldsymbol{\xi}) = a_{100}^{p_1 p_2 p_3} \omega_{100}(|\boldsymbol{\xi}|) \xi_1 + a_{010}^{p_1 p_2 p_3} \omega_{010}(|\boldsymbol{\xi}|) \xi_2 + a_{001}^{p_1 p_2 p_3} \omega_{001}(|\boldsymbol{\xi}|) \xi_3 + a_{200}^{p_1 p_2 p_3} \omega_{200}(|\boldsymbol{\xi}|) \xi_1^2$$
$$+ a_{020}^{p_1 p_2 p_3} \omega_{020}(|\boldsymbol{\xi}|) \xi_2^2 + a_{002}^{p_1 p_2 p_3} \omega_{002}(|\boldsymbol{\xi}|) \xi_3^2 + a_{110}^{p_1 p_2 p_3} \omega_{110}(|\boldsymbol{\xi}|) \xi_1 \xi_2 + a_{101}^{p_1 p_2 p_3} \omega_{101}(|\boldsymbol{\xi}|) \xi_1 \xi_3$$
$$+ a_{011}^{p_1 p_2 p_3} \omega_{011}(|\boldsymbol{\xi}|) \xi_2 \xi_3 \tag{28}$$

The weight function, $\omega_{q_1 q_2 q_3}(|\boldsymbol{\xi}|)$ can be expressed as:

$$\omega_{q_1 q_2 q_3}(|\boldsymbol{\xi}|) = \left(\frac{\delta}{|\boldsymbol{\xi}|} \right)^{q_1 + q_2 + q_3 + 1} \tag{29}$$

where $q_1, q_2, q_3 = 0, 1, 2$ and they cannot be equal to zero at the same time.

According to Equation (28), the peridynamic functions, $g_3^{p_1 p_2 p_3}(\boldsymbol{\xi})$, are composed of elements with an unknown coefficient matrix, **a**, weight function, ω and bond length in the x, y and z directions.

The unknown coefficient matrix, **a**, is derived from the peridynamic shape matrix, **A**, and the known coefficient matrix, **b**, via following relationship:

$$\mathbf{A}\mathbf{a} = \mathbf{b} \tag{30}$$

The elements of the peridynamic shape matrix, **A**, can be expressed as:

$$A_{(n_1 n_2 n_3)(q_1 q_2 q_3)} = \int_{Hx} \omega_{q_1 q_2 q_3}(|\boldsymbol{\xi}|) \xi_1^{n_1 + q_1} \xi_2^{n_2 + q_2} \xi_3^{n_3 + q_3} dV \tag{31}$$

where the unit volume is $dV_{\mathbf{x}'} = \xi^2 \sin \theta d\xi d\theta d\phi$. Hence, the peridynamic shape matrix can be calculated as:

$$
\mathbf{A} =
\begin{bmatrix}
\frac{4}{9}\pi\delta^5 & 0 & 0 & 0 & 0 & 0 & 0 & 0 & 0 \\
0 & \frac{4}{9}\pi\delta^5 & 0 & 0 & 0 & 0 & 0 & 0 & 0 \\
0 & 0 & \frac{4}{9}\pi\delta^5 & 0 & 0 & 0 & 0 & 0 & 0 \\
0 & 0 & 0 & \frac{1}{5}\pi\delta^7 & \frac{1}{15}\pi\delta^7 & \frac{1}{15}\pi\delta^7 & 0 & 0 & 0 \\
0 & 0 & 0 & \frac{1}{15}\pi\delta^7 & \frac{1}{5}\pi\delta^7 & \frac{1}{15}\pi\delta^7 & 0 & 0 & 0 \\
0 & 0 & 0 & \frac{1}{15}\pi\delta^7 & \frac{1}{15}\pi\delta^7 & \frac{1}{5}\pi\delta^7 & 0 & 0 & 0 \\
0 & 0 & 0 & 0 & 0 & 0 & \frac{1}{15}\pi\delta^7 & 0 & 0 \\
0 & 0 & 0 & 0 & 0 & 0 & 0 & \frac{1}{15}\pi\delta^7 & 0 \\
0 & 0 & 0 & 0 & 0 & 0 & 0 & 0 & \frac{1}{15}\pi\delta^7
\end{bmatrix}.
\tag{32}
$$

The elements of the known matrix, **b**, can be expressed as:

$$
b^{p_1 p_2 p_3}_{n_1 n_2 n_3} = n_1! n_2! n_3! \delta_{n_1 p_1} \delta_{n_2 p_2} \delta_{n_3 p_3}
\tag{33}
$$

Hence, the known coefficient matrix **b** can be written as:

$$
\mathbf{b} =
\begin{bmatrix}
1 & 0 & 0 & 0 & 0 & 0 & 0 & 0 & 0 \\
0 & 1 & 0 & 0 & 0 & 0 & 0 & 0 & 0 \\
0 & 0 & 1 & 0 & 0 & 0 & 0 & 0 & 0 \\
0 & 0 & 0 & 2 & 0 & 0 & 0 & 0 & 0 \\
0 & 0 & 0 & 0 & 2 & 0 & 0 & 0 & 0 \\
0 & 0 & 0 & 0 & 0 & 2 & 0 & 0 & 0 \\
0 & 0 & 0 & 0 & 0 & 0 & 1 & 0 & 0 \\
0 & 0 & 0 & 0 & 0 & 0 & 0 & 1 & 0 \\
0 & 0 & 0 & 0 & 0 & 0 & 0 & 0 & 1
\end{bmatrix}
\tag{34}
$$

According to Equation (30), the unknown matrix **a** can be computed as:

$$
\mathbf{a} =
\begin{bmatrix}
\frac{9}{4\pi\delta^5} & 0 & 0 & 0 & 0 & 0 & 0 & 0 & 0 \\
0 & \frac{9}{4\pi\delta^5} & 0 & 0 & 0 & 0 & 0 & 0 & 0 \\
0 & 0 & \frac{9}{4\pi\delta^5} & 0 & 0 & 0 & 0 & 0 & 0 \\
0 & 0 & 0 & \frac{12}{\pi\delta^7} & \frac{-3}{\pi\delta^7} & \frac{-3}{\pi\delta^7} & 0 & 0 & 0 \\
0 & 0 & 0 & \frac{-3}{\pi\delta^7} & \frac{12}{\pi\delta^7} & \frac{-3}{\pi\delta^7} & 0 & 0 & 0 \\
0 & 0 & 0 & \frac{-3}{\pi\delta^7} & \frac{-3}{\pi\delta^7} & \frac{12}{\pi\delta^7} & 0 & 0 & 0 \\
0 & 0 & 0 & 0 & 0 & 0 & \frac{15}{\pi\delta^7} & 0 & 0 \\
0 & 0 & 0 & 0 & 0 & 0 & 0 & \frac{15}{\pi\delta^7} & 0 \\
0 & 0 & 0 & 0 & 0 & 0 & 0 & 0 & \frac{15}{\pi\delta^7}
\end{bmatrix}
\tag{35}
$$

Substituting Equations (29) and (35) back to Equation (28), the peridynamic functions $g_3^{p_1 p_2 p_3}(\boldsymbol{\xi})$ can be calculated as:

$$
g_3^{100} = \frac{9}{4\pi\delta^3} \frac{1}{|\boldsymbol{\xi}|} \sin\theta \cos\phi
\tag{36}
$$

$$
g_3^{010} = \frac{9}{4\pi\delta^3} \frac{1}{|\boldsymbol{\xi}|} \sin\theta \sin\phi
\tag{37}
$$

$$
g_3^{001} = \frac{9}{4\pi\delta^3} \frac{1}{|\boldsymbol{\xi}|} \cos\theta
\tag{38}
$$

$$
g_3^{200} = \frac{12}{\pi\delta^4} \frac{1}{|\boldsymbol{\xi}|} \sin^2\theta \cos^2\phi - \frac{3}{\pi\delta^4} \frac{1}{|\boldsymbol{\xi}|} \sin^2\theta \sin^2\phi - \frac{3}{\pi\delta^4} \frac{1}{|\boldsymbol{\xi}|} \cos^2\theta
\tag{39}
$$

$$
g_3^{020} = \frac{-3}{\pi\delta^4} \frac{1}{|\boldsymbol{\xi}|} \sin^2\theta \cos^2\phi + \frac{12}{\pi\delta^4} \frac{1}{|\boldsymbol{\xi}|} \sin^2\theta \sin^2\phi - \frac{3}{\pi\delta^4} \frac{1}{|\boldsymbol{\xi}|} \cos^2\theta
\tag{40}
$$

$$g_3^{002} = \frac{-3}{\pi\delta^4} \frac{1}{|\xi|} \sin^2\theta\cos^2\phi - \frac{3}{\pi\delta^4} \frac{1}{|\xi|} \sin^2\theta\sin^2\phi + \frac{12}{\pi\delta^4} \frac{1}{|\xi|} \cos^2\theta \tag{41}$$

$$g_3^{110} = \frac{15}{\pi\delta^4} \frac{1}{|\xi|} \sin^2\theta\cos\phi\sin\phi \tag{42}$$

$$g_3^{101} = \frac{15}{\pi\delta^4} \frac{1}{|\xi|} \sin\theta\cos\theta\cos\phi \tag{43}$$

$$g_3^{011} = \frac{15}{\pi\delta^4} \frac{1}{|\xi|} \sin\theta\cos\theta\sin\phi \tag{44}$$

where θ represents the angle between the projection of the bond over the x-y plane and x-axis. ϕ shows the angle between the bond and the z-axis, as shown in Figure 2 (red line shows the bond).

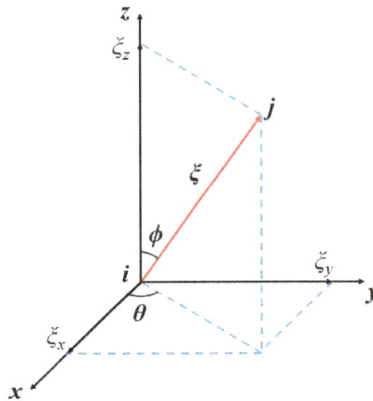

Figure 2. Diagram of three-dimensional bond orientation.

Substituting Equations (36)–(44) in Equation (27), the partial derivative expressions in peridynamic form can be obtained. Therefore, the extension of Fick's Second Law given in Equation (9) can be rewritten in peridynamic form as:

$$
\begin{aligned}
\frac{\partial C}{\partial t} = {}& Mk_B T \int_{H_x} \{C(\mathbf{x}',t) - C(\mathbf{x},t)\}\{g_3^{200}(\xi) + g_3^{020}(\xi) + g_3^{002}(\xi)\}dV_{\mathbf{x}'}\\
& - \frac{MC\Omega}{N_A}\int_{H_x} \{\tilde{\sigma}(\mathbf{x}',t) - \tilde{\sigma}(\mathbf{x},t)\}\{g_3^{200}(\xi) + g_3^{020}(\xi) + g_3^{002}(\xi)\}dV_{\mathbf{x}'}\\
& - \frac{MC\Omega}{N_A}\left\{\left(\int_{H_x}\{C(\mathbf{x}',t) - C(\mathbf{x},t)\}g_3^{100}(\xi)dV_{\mathbf{x}'}\right)\left(\int_{H_x}\{\tilde{\sigma}(\mathbf{x}',t) - \tilde{\sigma}(\mathbf{x},t)\}g_3^{100}(\xi)dV_{\mathbf{x}'}\right)\right\}\\
& - \frac{MC\Omega}{N_A}\left\{\left(\int_{H_x}\{C(\mathbf{x}',t) - C(\mathbf{x},t)\}g_3^{010}(\xi)dV_{\mathbf{x}'}\right)\left(\int_{H_x}\{\tilde{\sigma}(\mathbf{x}',t) - \tilde{\sigma}(\mathbf{x},t)\}g_3^{010}(\xi)dV_{\mathbf{x}'}\right)\right\}\\
& - \frac{MC\Omega}{N_A}\left\{\left(\int_{H_x}\{C(\mathbf{x}',t) - C(\mathbf{x},t)\}g_3^{001}(\xi)dV_{\mathbf{x}'}\right)\left(\int_{H_x}\{\tilde{\sigma}(\mathbf{x}',t) - \tilde{\sigma}(\mathbf{x},t)\}g_3^{001}(\xi)dV_{\mathbf{x}'}\right)\right\}
\end{aligned}
\tag{45}
$$

5. Numerical Studies

To demonstrate the capability of the current approach, spherical and cylindrical shaped anodes with different crack configurations were considered and the evolution of the damage during the lithiation process was investigated. In all cases, for the numerical solution, uniform spatial discretization was utilized with a grid size of 20 nm for the spherical model and 10 nm for the cylindrical model. Corresponding horizon sizes were specified as 60 nm and 30 nm, respectively. However, the recently introduced dual-horizon concept [20,22] can also be applied to improve the numerical efficiency of the solution. For the time integration, two different critical time step sizes for Equations (10) and (45) were determined as given in [34]. Since the critical time step size for each equation was different, we chose the smaller value, i.e., 2×10^{-14} s, to ensure the stability of the solution.

5.1. Spherical Model

In the first case, a spherical energy storage particle with a pre-existing penny-shaped crack was considered. The materials of this particle, including pure silicon and lithiated silicon, were regarded as brittle materials. In addition, the concentration values were normalized by the maximum concentration as shown in Table 1. Before the charging started, the anode material remained as pure silicon. During the charging process, lithium ions at their maximum concentration were applied to the entire outer surface. Since the particle structure was free from any displacement constraints, it experienced free expansion during the charging process. For the three-dimensional bond-based peridynamic theory, the Poisson's ratio is limited to 1/4. However, the Poisson's ratio values of pure silicon and lithiated silicon, given in Table 1, are very close to the constrained value. Therefore, the limitation of bond-based peridynamic theory regarding Poisson's ratio would not be effective in this case. Other Poisson's ratio values can be specified using the ordinary-state based peridynamic formulation. Since the geometry of the spherical energy storage particle was symmetric and the penny-shaped crack was horizontally oriented, the sample planes were selected along the longitude and latitude of the sphere, shown in red and black, respectively, in Figure 3 to show the lithium ion concentration and mechanical deformation clearly inside the particle.

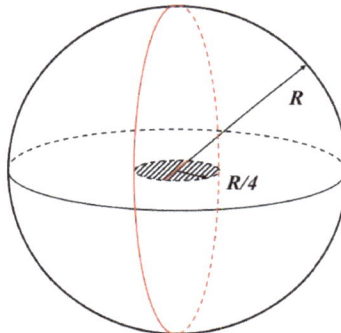

Figure 3. Spherical energy storage particle with a pre-existing penny-shaped crack.

Table 1. Geometrical parameters and material properties of energy storage models.

R	Radius of spherical particle	0.5 μm
L	Longitudinal length of cylindrical nanowire	0.5 μm
$E_{silicon}$	Elastic constant of pure silicon	80 GPa
$E_{Li_{15}Si_4}$	Elastic constant of lithiated silicon, $Li_{15}Si_4$	41 GPa
$\nu_{silicon}$	Poisson's ratio of pure silicon	0.22
$\nu_{Li_{15}Si_4}$	Poisson's ratio of lithiated silicon, $Li_{15}Si_4$	0.24
Ω	Partial molar volume	8.5×10^{-6} m^3 mol^{-1}
M	Molecular mobility	500 m^2 J^{-1} s^{-1}
k_B	Boltzmann constant	1.38×10^{-23} J·K^{-1}
T	Absolute temperature	300 K
N_A	Avogadro's constant	6.02×10^{23} mol^{-1}
C_{max}	Maximum concentration	1.18×10^4 mol m^{-3}

The penny-shaped crack was located in the central region of the anode structure. The diameter of the penny-shaped crack was half of the diameter of the anode and the crack was horizontally oriented, as shown in Figure 3. As lithiation progressed, the material at the anode surface region became lithiated silicon ($Li_{15}Si_4$) first. Hence, the surface region experienced relatively large deformation while the central region remained undeformed. Due to this mechanical deformation state, compressive stresses emerged at the particle surface region, whereas tensile stresses formed at the particle central region, especially around the crack tip location. Hence, according to Equation (9), high hydrostatic stress affects the lithium ion distribution inside the spherical structure.

The results of lithium ion diffusion and diffusion-induced deformation during the charging process are shown in Figures 4 and 5. In order to have a clear understanding of the lithiation-induced damage in the three-dimensional structure, results are presented in both the x-z plane and x-y plane (crack surface plane). Due to the heterogeneous distribution of deformation in the spherical structure, high hydrostatic stresses arose at the edge of the penny-shaped crack where high geometrical singularity lies, as shown in Figures 4c and 5c. As the lithium ion diffused further into the spherical structure, the lithium ion concentration at the crack edge regions started to increase compared with the surrounding regions. However, since the penny-shaped crack had just started to propagate, as shown in Figures 4b and 5b, the change in lithium-ion concentration may not be very obvious from the concentration plot. The hydrostatic stress value around the crack edge region was around 1.5 GPa which led to a bond stretch far beyond the critical stretch of amorphous silicon. Hence, the penny-shaped crack would have continued to propagate until the stretch of the bonds at the crack tip region reduced below the critical stretch value given in Equation (16).

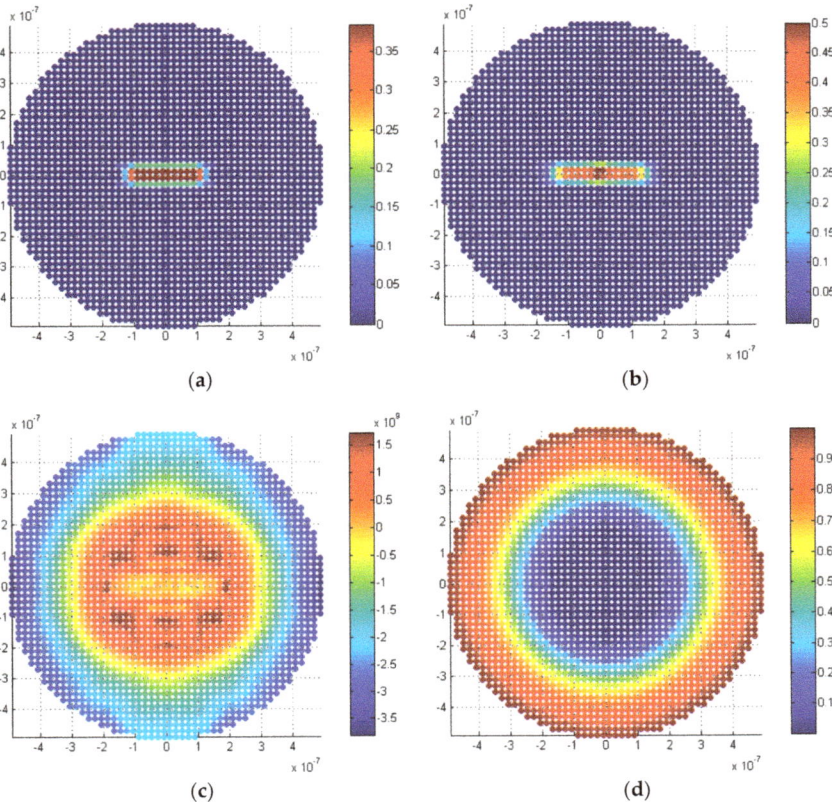

Figure 4. Results for a penny-shaped, cracked anode in the x-z mid-plane: (**a**) initial damage; (**b**) new damage; (**c**) hydrostatic stress; and (**d**) lithium ion concentration.

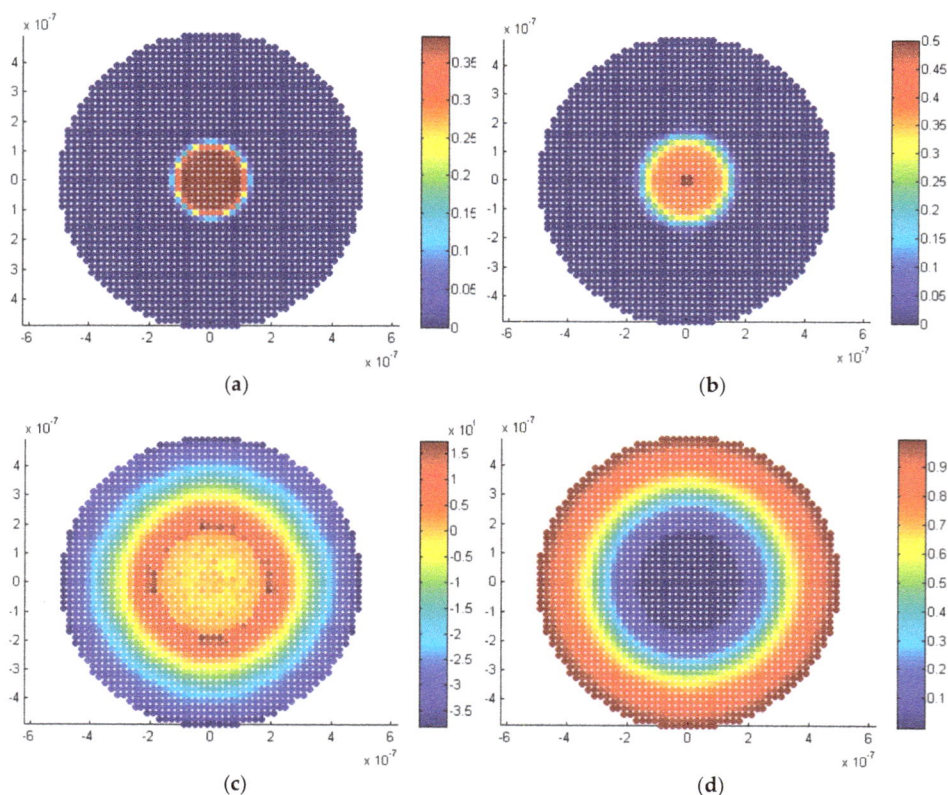

Figure 5. Results for a penny-shaped, cracked anode in the *x-y* mid-plane (crack surface plane): (**a**) initial damage; (**b**) new damage; (**c**) hydrostatic stress; and (**d**) lithium ion concentration.

5.2. Cylindrical Model

In this section, several different pre-damaged models are considered to investigate the fracture evolution in the silicon nanowire. The silicon nanowire is represented by a cylindrical-shaped structure. The fracture analysis of cases with different crack orientations is separately discussed in the following sections.

5.2.1. Penny-Shaped Crack along the Horizontal Plane

A cylindrical-shaped structure with diameter L, shown in Figure 6, was considered to represent a silicon nanowire. A penny-shaped crack (marked as shadow lines in Figure 6), $L/4$ in diameter, was horizontally oriented at the center of the cylindrical structure. The material properties of the cylindrical anode were the same as for the spherical model in Table 1. Before charging the battery, the cylindrical material was pure silicon. During the charging process, lithium ions at maximum concentration was applied to the outer surface of the cylindrical structure. As a result, the material properties changed as the lithium ions diffused inside the electrode and the cylindrical structure deformed. Similar to the spherical model, the results are shown in two different plane views in order to present detailed fracture information.

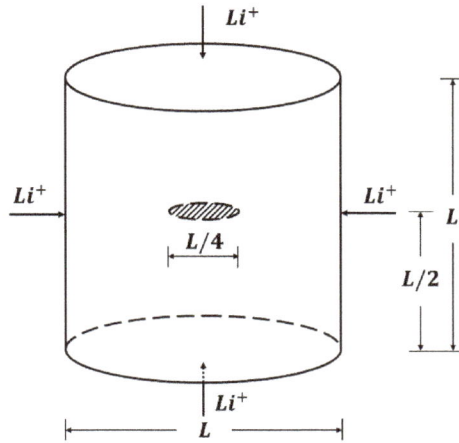

Figure 6. Cylindrical anode structure with a pre-existing horizontal crack.

Due to the material phase change during the charging process, material points at the surface region expanded as pure silicon changed into lithiated silicon, whereas material points in the central region remained undeformed. Therefore, the hydrostatic stress in the outer surface region is shown as compression stress which is marked as blue in Figures 7c and 8c. The hydrostatic stress in the inner region of the cylindrical anode is shown as tensile stress, which is marked in red, especially at the crack tip region. In this case, the crack propagation is obvious. The penny-shaped crack edge lies close to the material phase boundary, as shown Figures 7b and 8b. Hence, according to Equation (9), the lithium ion concentration at the crack edge region is relatively higher than in the surrounding regions as shown in Figures 7d and 8d. Since the hydrostatic stress at the crack tip region is around 2.3 GPa, it leads to the bond stretching far beyond the critical stretch of the lithiated silicon. Therefore, the crack propagation will not stop until the stretch values of the bonds at the crack edge regions reduce below the critical stretch value.

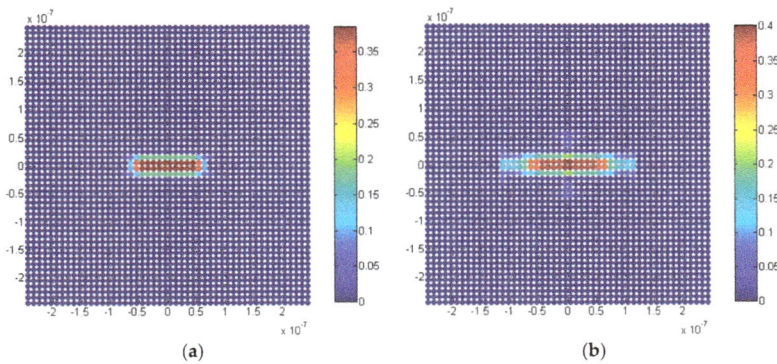

Figure 7. *Cont.*

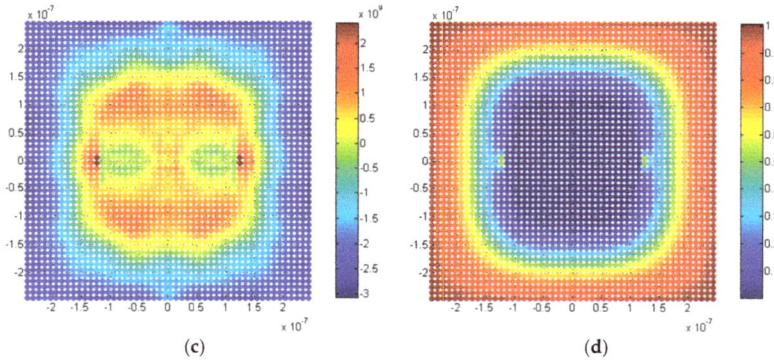

Figure 7. Result plots of the cylindrical anode with a single horizontal crack in the *x*-*z* mid-plane: (**a**) initial damage; (**b**) damage after deformation; (**c**) hydrostatic stress; and (**d**) lithium ion concentration.

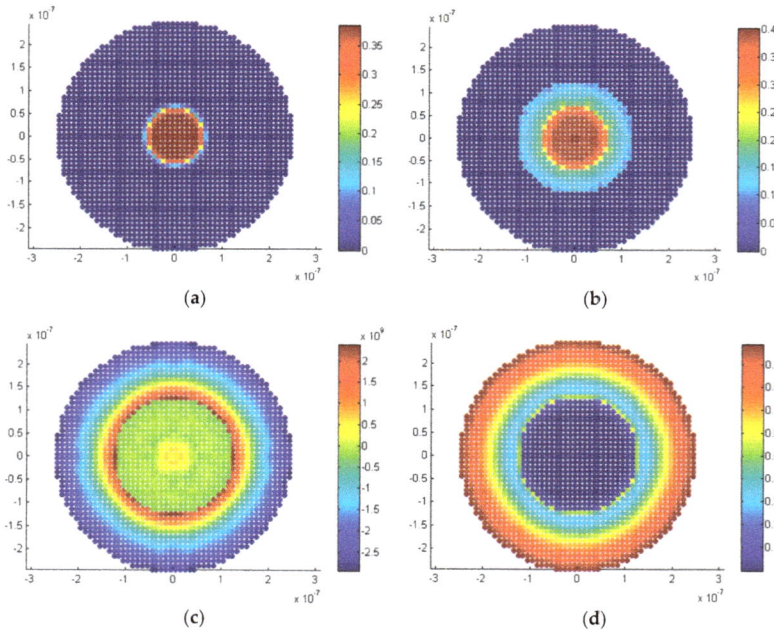

Figure 8. Result plots of the cylindrical anode with a single horizontal crack in the *x*-*y* mid-plane (crack surface plane): (**a**) initial damage; (**b**) damage after deformation; (**c**) hydrostatic stress; and (**d**) lithium ion concentration.

5.2.2. Penny-Shaped Crack along the Vertical Plane

In contrast to the spherical energy storage particle model presented earlier, the cylindrical silicon nanowire model may show different fracture behaviors in the radial direction and longitudinal direction. Hence, in this case study, the penny-shaped crack along the longitudinal (*z*-direction) direction is under investigation. A penny-shaped crack with a diameter of $L/4$ was located at the central position in the cylinder model, as shown in Figure 9. However, in contrast to the case study in Section 5.2.1, the crack was vertically oriented (or along the *x-z* mid-plane). Before charging the lithium ion battery, the cylinder was composed of pure silicon and the lithium ion concentration was zero throughout the whole structure. During the charging process, the lithium ion concentration at maximum value was applied to the entire outer surface of the cylinder. As the lithium ion concentration increased, pure silicon transformed into fully lithiated silicon which led to volume expansion. The stress and damage induced by this volume expansion are shown in Figures 9 and 10.

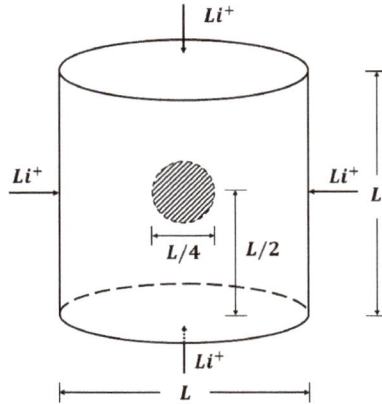

Figure 9. Cylindrical anode structure with a pre-existing vertical crack.

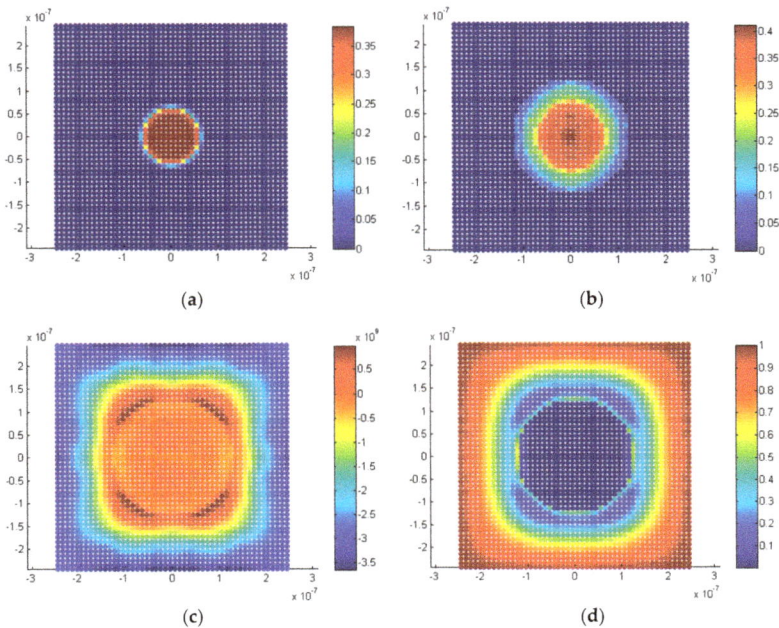

Figure 10. Result plots of cylindrical anode with a single vertical crack in the *y-z* mid-plane: (**a**) initial damage; (**b**) damage after deformation; (**c**) hydrostatic stress; and (**d**) lithium ion concentration.

The results are presented in the *x-z* mid-plane view (Figure 10) and *x-y* mid-plane view (Figure 11). During the charging process, the material points in the outer surface region were first lithiated and then expanded whereas the material points in the central region remained undeformed. As a result, a deformation gradient from outer surface to central cylinder was produced, and hydrostatic stress increased, especially at the crack edge regions, in accordance with Equations (19)–(23). Figures 10c and 11c, show the location of compression stress at the outer surface region, which is represented in blue in the figure. Tension stress exists in the central cylinder region, especially at crack edge regions where geometrical singularity exists, which is represented as red in the figure. As the crack propagated, the crack opened which caused relatively high tension stress near the crack surface. Since the crack propagated close to the material phase boundary, the lithium ion concentration was relatively higher than surrounding regions in accordance with to Equation (9), as shown in Figures 10d and 11d.

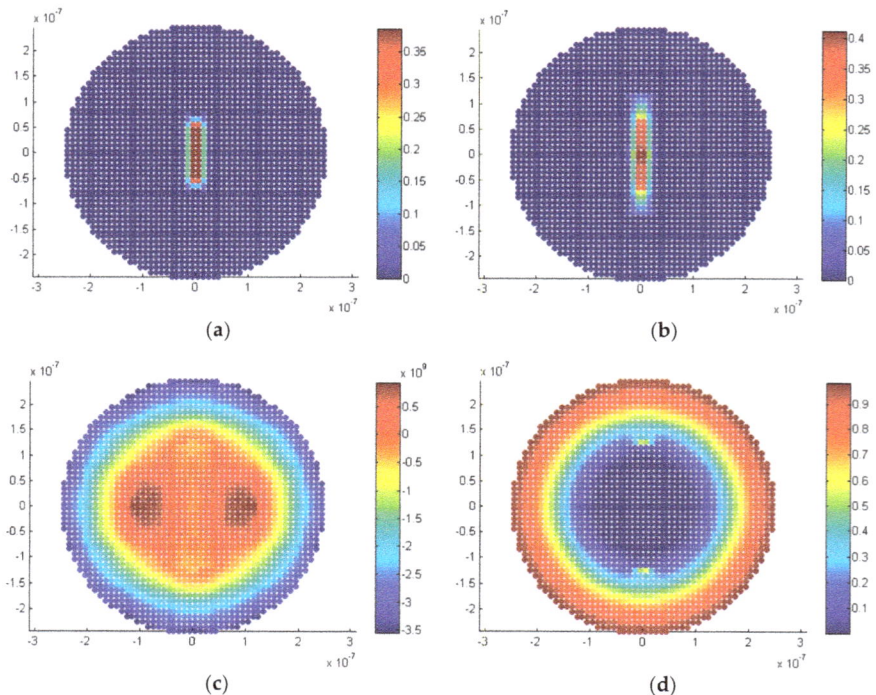

Figure 11. Result plots of cylindrical anode with a single vertical crack in the *x-y* mid-plane: (**a**) initial damage; (**b**) damage after deformation; (**c**) hydrostatic stress; and (**d**) lithium ion concentration.

5.2.3. Twin Penny-Shaped Cracks along Horizontal Planes

The case studies in Sections 5.2.1 and 5.2.2 focused on the fracture analysis of a single crack on the symmetry plane of the cylinder structure. Hence, due to symmetric geometry and loading, the penny-shaped crack propagated along the crack surface plane. However, in reality, multiple cracks can occur at any position inside the battery electrode. Hence, in this case study, a silicon cylinder with two cracks not on the symmetry plane was investigated. As shown in Figure 12, twin cracks were located at the central region with diameters equal to $L/4$ and the distance between these cracks was also $L/4$. Both of the cracks were horizontally oriented. Before charging the battery, the cylindrical nanowire was composed of pure silicon only. During the charging process, lithium ions at maximum concentration were applied on the entire outer surface of the cylinder. As a result, the material points on the surface became lithiated and expanded while the material points in the central region remained undeformed. The heterogeneous deformation led to an increase in hydrostatic stress in regions with geometrical singularity and sped up the diffusion process, as shown in Figure 13.

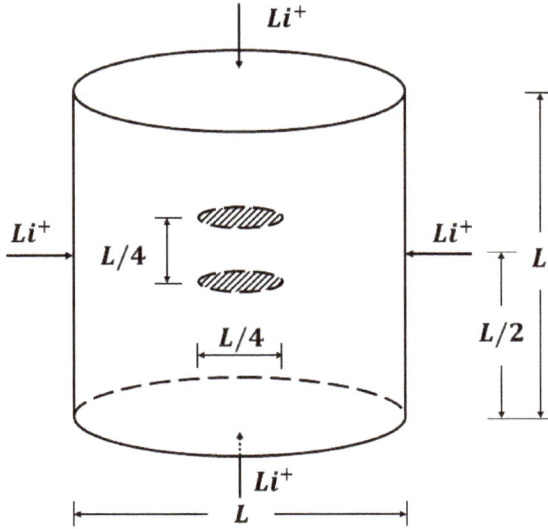

Figure 12. Cylindrical anode structure with pre-existing twin horizontal cracks.

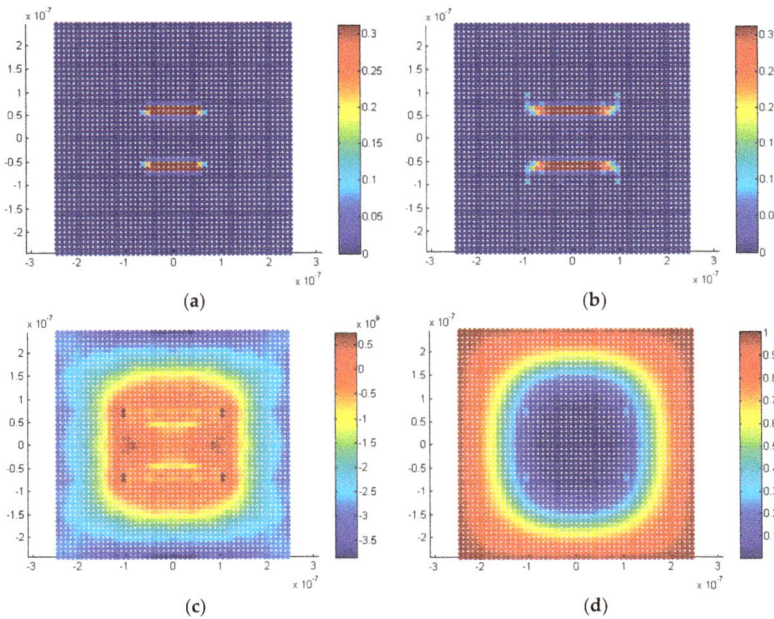

Figure 13. Result plots of a silicon cylinder with twin horizontal cracks in the *x-z* mid-plane: (**a**) initial damage; (**b**) damage after deformation; (**c**) hydrostatic stress; and (**d**) lithium ion concentration.

Compared with the case in Section 5.2.1, the crack propagation behavior was different. At the initial stage of the charging process, both cracks propagated along the crack surface plane. However, as the charging process continued, the upper crack propagated upward from its edge and the lower crack propagated downwards from its edge. These cracks repelled each other, as shown in Figure 13b. Due to the heterogeneous volume expansion inside the cylinder, a deformation gradient formed from surface of the cylinder to the central cylinder. As a result, the hydrostatic stress increased, especially at the crack edge regions, in accordance with Equations (19)–(23). Since material points at the surface region experienced expansion during the charging process, the compression stresses were evident, as represented in Figure 13c. The expansion of these material points caused tensile stress for the material points in the central region, especially around the crack edge region, as shown in Figure 13c. Since the crack edge is close to the material phase boundary, lithium ions would have diffused into regions around the crack edge with higher priority in accordance with Equation (9). Hence, the lithium ion concentration was relatively higher than surrounding regions, as shown in Figure 13d.

5.2.4. Twin Penny Shape Cracks along Vertical Planes

After the analysis of twin cracks along horizontal planes, a case with twin cracks along vertical planes was also considered. In this case study, twin vertical oriented cracks were arranged in the center region of a silicon cylinder, as shown in Figure 14. The diameters of both cracks were $L/4$ and the distance between these cracks was also $L/4$. Before charging the battery, the silicon cylinder was made of pure silicon. During the charging process, lithium ions at maximum concentration were applied on the entire outer surface of the cylinder. Hence, the material particles inside the outer surface region became lithiated and expanded. This influenced the lithium ion diffusion and hydrostatic stress inside the cylinder, as shown in Figure 15.

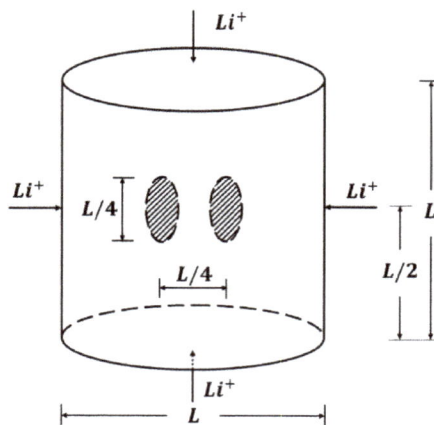

Figure 14. Cylindrical anode structure with pre-existing twin vertical cracks.

Figure 15a shows the initial damage in the view of the x-z mid-plane. Early in the charging process, penny shape cracks propagated along their crack surfaces. However, as the charging process continued, the left crack propagated towards the left and the right crack propagated towards the right. Generally, these cracks repelled each other, as shown in Figure 15b. Heterogeneous volume expansion led to the formation of a deformation gradient from the surface region to the central region. Therefore, hydrostatic stress rose inside the cylinder in accordance with Equations (19)–(23). The heterogeneous expansion during lithiation introduced compression stress on material points at the cylinder surface region, which is marked in blue in Figure 15c. As a consequence, the material points at the central cylinder region would have suffered from the tensile stress, marked in red. As the crack edge reached the material phase boundary, the lithium ion would have diffused into material points in the crack edge region first. Hence, a relatively high lithium ion concentration at the crack edge region as compared with its surrounding region was observed, as shown in Figure 15d.

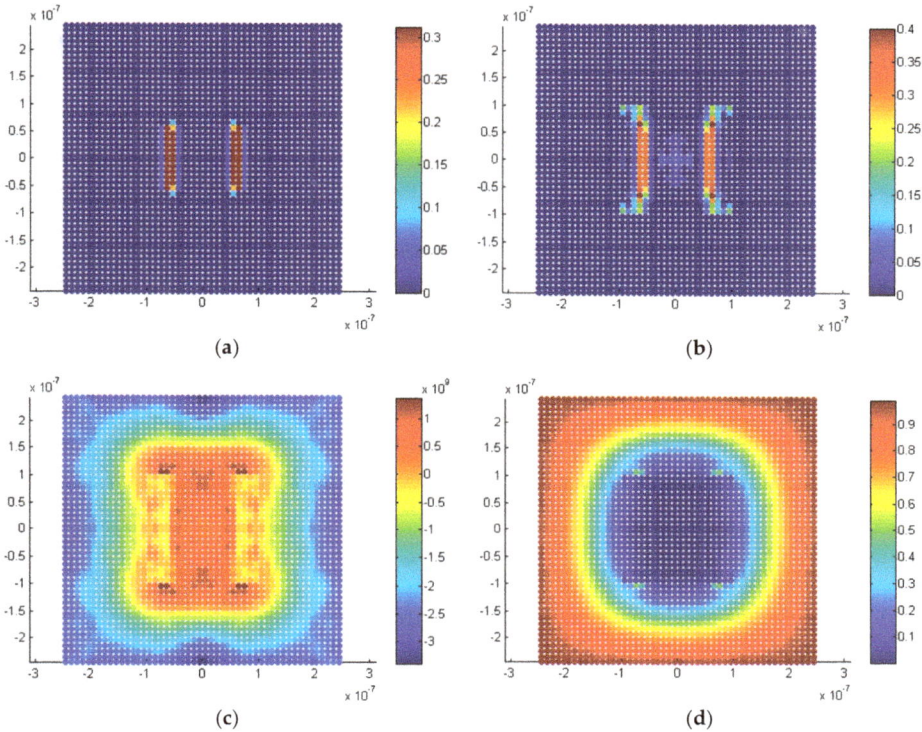

Figure 15. Result plots of twin vertical cracks cylinder in the *x-z* mid-plane: (**a**) initial damage; (**b**) damage after deformation; (**c**) hydrostatic stress; and (**d**) lithium ion concentration.

5.3. Discussion

The results presented show that lithiation can influence the fracture behavior of a battery electrode. In the perspective of peridynamic theory, crack propagation is calculated based on the bond stretch which depends on the deformation of the structure. Hence, crack propagation generally depends on deformation which is induced by lithiation during battery charging. Since the pure silicon transforms into lithiated silicon, the material properties also change which leads to change in the critical bond stretch value. The comparison between fracture behavior with and without material phase change has been discussed in previous works [26,27]. In this article, since the material phase change during the charging process was also under consideration, the critical bond stretch was also influenced by the material properties.

For an electrode structure with symmetric geometry, the method of crack propagation is different depending on the position of the initial crack. If the initial cracks lie on the symmetry plane, cracks will propagate along the crack surface plane, since the geometry and loading are symmetric. However, if the initial cracks do not lie on the symmetry plane, the newly formed cracks may not align with the initial crack. For the twin cracks cases, the two-dimensional penny-shaped cracks turned into three-dimensional bowl-shaped cracks after the battery as charged. This is because of the material softening phenomenon during lithiation, as described in [32]. Due to the increase in lithium ion concentration, pure silicon at the crack edge regions will transform into lithiated silicon ($Li_{15}Si_4$). As a consequence, the critical bond stretch is reduced. Hence, the bonds at crack edge region reach a critical stretch value earlier which leads the crack edge to propagate towards the high lithium ion concentration region. High hydrostatic stresses exist between the twin cracks, as shown in Figures 13c and 15c, and the bonds located in these regions may not reach the critical value. However, in situations of high hydrostatic stress, there is potential for these two cracks to merge into one larger crack as shown in Figure 15b.

6. Conclusions

In this study, peridynamic theory, in conjunction with the newly developed three-dimensional peridynamic differential operator approach, was utilized as a new method in three-dimensional fracture analysis during the lithiation process. Two differently shaped structures, a spherical energy storage particle and cylindrical nanowire, with different pre-existing cracks, were considered. According to numerical results, crack propagation is usually influenced by high hydrostatic stresses which yield bond breakage. In addition, the material phase change from pure silicon to lithiated silicon also influences the crack propagation. If the model geometry and loading are symmetric, the direction of crack propagation also depends on the initial crack position. High hydrostatic stresses lie between the cracks in twin crack cases, which means twin cracks have a potential to merge into one larger crack.

As a summary, it was shown that peridynamic theory is a suitable candidate for prediction of three-dimensional damage evolution in lithium ion battery electrodes. In peridynamic theory, the damage formation and evolution can be simplified without remeshing or applying sophisticated damage criteria. By using peridynamics, one can have better understanding of the failure mechanisms during the cyclic operation of lithium ion batteries.

Author Contributions: H.W., E.O. and S.O. contributed to the design and implementation of the research. H.W. and E.O. wrote the manuscript.

Funding: This research received no external funding.

Conflicts of Interest: The authors declare no conflict of interest.

References

1. Annex, V.I. *Regulations for the Prevention of Air Pollution from Ships and NOx Technical Code*; MARPOL 73/78; International Maritime Organization (IMO): London, UK, 1998.
2. Rosenkranz, C. Deep-cycle batteries for plug-in hybrid application. In Proceedings of the EVS20 Plug-In Hybrid Vehicle Workshop, Long Beach, CA, USA, 15–19 November 2003.
3. Chung, S.Y.; Bloking, J.T.; Chiang, Y.M. Electronically conductive phospho-olivines as lithium storage electrodes. *Nat. Mater.* **2002**, *1*, 123–128. [CrossRef] [PubMed]
4. De las Casas, C.; Li, W. A review of application of carbon nanotubes for lithium ion battery anode material. *J. Power Sources* **2012**, *208*, 74–85. [CrossRef]
5. Zhang, W.J. A review of the electrochemical performance of alloy anodes for lithium-ion batteries. *J. Power Sources* **2011**, *196*, 13–24. [CrossRef]
6. Nadimpalli, S.P.V.; Tripuraneni, R.; Sethuraman, V.A. Real-time stress measurements in germanium thin film electrodes during electrochemical Lithiation/Delithiation cycling. *J. Electrochem. Soc.* **2015**, *162*, A2840–A2846. [CrossRef]
7. Chan, C.K.; Peng, H.; Liu, G.; McIlwrath, K.; Zhang, X.F.; Huggins, R.A.; Cui, Y. High-performance lithium battery anodes using silicon nanowires. *Nat. Nanotechnol.* **2008**, *4*, 31–35. [CrossRef] [PubMed]
8. Ryu, I.; Choi, J.W.; Cui, Y.; Nix, W.D. Size-dependent fracture of Si nanowire battery anodes. *J. Mech. Phys. Solids* **2011**, *4*, 1717–1730. [CrossRef]
9. Ren, J.G.; Wu, Q.H.; Tang, H.; Hong, G.; Zhang, W.; Lee, S.T. Germanium–graphene composite anode for high-energy lithium batteries with long cycle life. *J. Mater. Chem. A* **2013**, *4*, 1821–1826. [CrossRef]
10. Winter, M.; Besenhard, J.O.; Spahr, M.E.; Novak, P. Insertion electrode materials for rechargeable lithium batteries. *Adv. Mater.* **1998**, *4*, 725–763. [CrossRef]

11. Gao, B.; Kleinhammes, A.; Tang, X.P.; Bower, C.; Fleming, L.; Wu, Y.; Zhou, O. Electrochemical intercalation of single-walled carbon nanotubes with lithium. *Chem. Phys. Lett.* **1999**, *307*, 153–157. [CrossRef]
12. Lee, J.H.; Kong, B.S.; Yang, S.B.; Jung, H.T. Fabrication of single-walled carbon nanotube/tin nanoparticle composites by electrochemical reduction combined with vacuum filtration and hybrid co-filtration for high-performance lithium battery electrodes. *J. Power Sources* **2009**, *4*, 520–525. [CrossRef]
13. Yan, J.; Song, H.; Yang, S.; Yan, J.; Chen, X. Preparation and electrochemical properties of composites of carbon nanotubes loaded with Ag and TiO_2 nanoparticle for use as anode material in lithium-ion batteries. *Electrochim. Acta* **2008**, *4*, 6351–6355. [CrossRef]
14. Liu, P.; Sridhar, N.; Zhang, Y.W. Lithiation-induced tensile stress and surface cracking in silicon thin film anode for rechargeable lithium battery. *J. Appl. Phys.* **2012**, *112*, 093507. [CrossRef]
15. Wang, X.; Shen, W.; Huang, X.; Zang, J.; Zhao, Y. Estimating the thickness of diffusive solid electrolyte interface. *Sci. China Phys. Mech. Astron.* **2017**, *60*, 064612. [CrossRef]
16. Grantab, R.; Shenoy, V.B. Pressure-gradient dependent diffusion and crack propagation in lithiated silicon nanowires. *J. Electrochem. Soc.* **2012**, *4*, A584–A591. [CrossRef]
17. Zuo, P.; Zhao, Y.P. A phase field model coupling lithium diffusion and stress evolution with crack propagation and application in lithium ion batteries. *Phys. Chem. Chem. Phys.* **2015**, *4*, 287–297. [CrossRef] [PubMed]
18. Gao, Y.F.; Zhou, M. Coupled mechano-diffusional driving forces for fracture in electrode materials. *J. Power Sources* **2013**, *230*, 176–193. [CrossRef]
19. Silling, S.A. Reformulation of elasticity theory for discontinuities and long-range forces. *J. Mech. Phys. Solids* **2000**, *4*, 175–209. [CrossRef]
20. Ren, H.; Zhuang, X.; Cai, Y.; Rabczuk, T. Dual-horizon peridynamics. *Int. J. Numer. Methods Eng.* **2016**, *108*, 1451–1476. [CrossRef]
21. Amani, J.; Oterkus, E.; Areias, P.; Zi, G.; Nguyen-Thoi, T.; Rabczuk, T. A non-ordinary state-based peridynamics formulation for thermoplastic fracture. *Int. J. Impact Eng.* **2016**, *87*, 83–94. [CrossRef]
22. Ren, H.; Zhuang, X.; Rabczuk, T. Dual-horizon peridynamics: A stable solution to varying horizons. *Comput. Methods Appl. Mech. Eng.* **2017**, *318*, 762–782. [CrossRef]
23. Diyaroglu, C.; Oterkus, E.; Madenci, E.; Rabczuk, T.; Siddiq, A. Peridynamic modeling of composite laminates under explosive loading. *Compos. Struct.* **2016**, *144*, 14–23. [CrossRef]
24. Diyaroglu, C.; Oterkus, E.; Oterkus, S.; Madenci, E. Peridynamics for bending of beams and plates with transverse shear deformation. *Int. J. Solids Struct.* **2015**, *69*, 152–168. [CrossRef]
25. Oterkus, E.; Madenci, E. Peridynamics for failure prediction in composites. In Proceedings of the 53rd AIAA/ASME/ASCE/AHS/ASC Structures, Structural Dynamics and Materials Conference, Honolulu, HI, USA, 23–26 April 2012; p. 1692.
26. Wang, H.; Oterkus, E.; Oterkus, S. Peridynamic modelling of fracture in marine lithium-ion batteries. *Ocean Eng.* **2018**, *151*, 257–267. [CrossRef]
27. Wang, H.; Oterkus, E.; Oterkus, S. Predicting fracture evolution during lithiation process using peridynamics. *Eng. Fract. Mech.* **2018**, *192*, 176–191. [CrossRef]
28. Fick, A. On liquid diffusion. *J. Membr. Sci.* **1995**, *4*, 33–38. [CrossRef]
29. Zhang, X.; Lee, S.W.; Lee, H.W.; Cui, Y.; Linder, C. A reaction-controlled diffusion model for the lithiation of silicon in lithium-ion batteries. *Extrem. Mech. Lett.* **2015**, *4*, 61–75. [CrossRef]
30. Shenoy, V.B.; Johari, P.; Qi, Y. Elastic softening of amorphous and crystalline Li–Si phases with increasing Li concentration: A first-principles study. *J. Power Sources* **2010**, *4*, 6825–6830. [CrossRef]
31. Zienkiewicz, O.C.; Taylor, R.L.; Zienkiewicz, O.C.; Taylor, R.L. *The Finite Element Method*; McGraw-Hill: London, UK, 1977; Volume 3.
32. Brebbia, C.A.; Telles, J.C.F.; Wrobel, L.C. *Boundary Element Techniques: Theory and Applications in Engineering*; Springer Science & Business Media: Berlin/Heidelberg, Germany, 1984.
33. Moës, N.; Belytschko, T. Extended finite element method for cohesive crack growth. *Eng. Fract. Mech.* **2002**, *4*, 813–833. [CrossRef]
34. Madenci, E.; Oterkus, E. *Peridynamic Theory and Its Applications*; Springer: New York, NY, USA, 2014.
35. Silling, S.A.; Askari, E. A meshfree method based on the peridynamic model of solid mechanics. *Comput. Struct.* **2005**, *4*, 1526–1535. [CrossRef]
36. Silling, S.A.; Epton, M.; Weckner, O.; Xu, J.; Askari, E. Peridynamic states and constitutive modeling. *J. Elast.* **2007**, *4*, 151–184. [CrossRef]

37. Oterkus, S.; Madenci, E.; Oterkus, E.; Hwang, Y.; Bae, J.; Han, S. Hygro-thermo-mechanical analysis and failure prediction in electronic packages by using peridynamics. In Proceedings of the 2014 IEEE 64th Electronic Components and Technology Conference (ECTC), Lake Buena Vista, FL, USA, 27–30 May 2014; pp. 973–982.

38. Rabczuk, T.; Belytschko, T. A three dimensional large deformation meshfree method for arbitrary evolving cracks. *Comput. Methods Appl. Mech. Eng.* **2007**, *196*, 2777–2799. [CrossRef]

39. Rabczuk, T.; Belytschko, T. Cracking particles: A simplified meshfree method for arbitrary evolving cracks. *Int. J. Numer. Methods Eng.* **2004**, *4*, 2316–2343. [CrossRef]

40. Malvern, L.E. *Introduction to the Mechanics of a Continuous Medium*; Prentice-Hall: Hertfordshire, UK, 1969.

41. Madenci, E. Peridynamic integrals for strain invariants of homogeneous deformation. *ZAMM J. Appl. Math. Mech./Z. Angew. Math. Mech.* **2017**, *4*, 1236–1251. [CrossRef]

42. Madenci, E.; Barut, A.; Futch, M. Peridynamic differential operator and its applications. *Comput. Methods Appl. Mech. Eng.* **2016**, *304*, 408–451. [CrossRef]

Article

Thermal Conductance along Hexagonal Boron Nitride and Graphene Grain Boundaries

Timon Rabczuk [1,2,*,†], **Mohammad Reza Azadi Kakavand** [3,†], **Raahul Palanivel Uma** [4],
Ali Hossein Nezhad Shirazi [5] and **Meysam Makaremi** [6]

[1] Division of Computational Mechanics, Ton Duc Thang University, Ho Chi Minh City, Vietnam
[2] Faculty of Civil Engineering, Ton Duc Thang University, Ho Chi Minh City, Vietnam
[3] Unit of Strength of Materials and Structural Analysis, Institute of Basic Sciences in Engineering Sciences, University of Innsbruck, 6020 Innsbruck, Austria; mohammad.azadi-kakavand@uibk.ac.at
[4] Institute of Mechanics, University of Duisburg-Essen, 45141 Essen, Germany; rahulpalani1990@gmail.com
[5] Institute of Structural Mechanics, Bauhaus-University of Weimar, 99423 Weimar, Germany; ali.hn.s@outlook.com
[6] Faculty of Applied Science & Engineering, University of Toronto, Toronto, ON M5S 3E4, Canada; meysam.makaremi@gmail.com
* Correspondences: timon.rabczuk@tdt.edu.vn
† These authors contributed equally to this work.

Received: 15 April 2018; Accepted: 11 June 2018; Published: 14 June 2018

Abstract: We carried out molecular dynamics simulations at various temperatures to predict the thermal conductivity and the thermal conductance of graphene and hexagonal boron-nitride (h-BN) thin films. Therefore, several models with six different grain boundary configurations ranging from 33–140 nm in length were generated. We compared our predicted thermal conductivity of pristine graphene and h-BN with previously conducted experimental data and obtained good agreement. Finally, we computed the thermal conductance of graphene and h-BN sheets for six different grain boundary configurations, five sheet lengths ranging from 33 to 140 nm and three temperatures (i.e., 300 K, 500 K and 700 K). The results show that the thermal conductance remains nearly constant with varying length and temperature for each grain boundary.

Keywords: molecular dynamics simulation; h-BN and Graphene sheets; thermal conductance; thermal conductivity

1. Introduction

Due to its exceptional material properties graphene is a promising material for numerous industrial applications ranging from nanoelectronics to the aerospace industry. As the strongest man-made material, graphene possesses a tensile strength of 130 ± 10 GPa, an elastic modulus of 1 ± 0.1 TPa, and also a high thermal conductivity of 4100 ± 500 W/m·K [1–6].

The discovery of graphene has motivated scientists to study other 2D materials such as hexagonal boron-nitride (h-BN), which also exhibits noteworthy mechanical, thermal, and electrical properties but in contrast to graphene, it possesses a natural band gap. The physical structure of graphene; i.e., a thin-layer of carbon atoms forming a honey-comb lattice structure—enables it to be used in applications such as hetero-structured photo detectors, strain sensors, and even composites combined with aluminum to improve heat-dissipation properties [7–10]. Possessing a similar physical structure as that of Graphene, h-BN shows very good mechanical, thermal insulating, and dielectric properties, in addition to chemical stability [11]. By using different combinations of stiffness, ultimate tensile strength and strain, the mechanical properties of h-BN can be tuned and used in advanced applications such as strain-engineered nano-ribbons, bio-compatible nano-devices and nano-electronics [12].

The heterogeneous integration of both materials, due to their similar physical structures, results in the increase of the overall tensile strength, thermal, and electrical conductivities, which is useful for developing semiconductor-insulators [13–19].

Thermal chemical vapor deposition (CVD) is the most common technique used to produce thin-film sheets of graphene and hexagonal boron-nitride. During the fabrication process, polycrystalline growth leads to grain boundaries and defects along the boundaries of the lattice surface. These defects are of different shapes; the most common type being pentagon-heptagon shaped. Less common shapes include square-octagon pairs. The pentagon-heptagon shaped pairs are comprised of homonuclear boron-boron or nitrogen-nitrogen pairs rather than heteronuclear boron-nitrogen pairs, which constitute the rest of the hexagonal lattice structure in the entire thin-film sheet. These defects act as dislocations and cause stress concentrations and phonon scattering of the grains reducing the tensile strength as well as the thermal and electrical conductivity of the materials, thereby affecting its structural stability during loading conditions [13–20].

Computational methods such as molecular dynamics (MD) have been employed to study the effect of grain boundaries. Defect patterns are modelled for symmetric and non-symmetric grain boundaries that occur at different angles depending on the armchair and zigzag orientation of the defect-patterns along the lattice structure. The number of grain boundaries increases with increasing misorientation angle along the boundaries of the polycrystalline lattice structure. This rise in the number of grain boundaries leads to an enhanced thermal and electrical resistance, which might affect their efficiency. Hence, a basic understanding of the grain boundaries and their impact on thermal, mechanical, and electrical properties is essential for potential industrial applications [21,22].

In this study, we carried out MD simulations in order to investigate the thermal properties of h-BN and graphene sheets containing ultra-fine grains up to 1 nm. The thermal properties were computed for different sheet dimensions in order to fully understand the effect of the grain boundaries on the surface of the lattice sheets and their influence on the heat flow across the surface of the materials.

2. Modeling Using Molecular Dynamics

In this study, we used our own C++ script to construct the required grain boundaries while the MD simulations were done with the LAMMPS package (Large-scale Atomic/Molecular Massively Parallel Simulator) [23]. Due to the combination of the planar geometry and its high thermo-mechanical and oxidation-resistant properties, hexagonal boron-nitride can be employed for several other industrial applications including high temperature membrane filters, as fillers in reinforced polymeric materials, and high oxidation-resistant coatings for metals such as nickel. Moreover, graphene may be employed as transparent conducting layer for high temperature thin film device applications where temperatures can reach up to 1100 °C [24–26]. Hence, the influence of various sheet lengths ranging from 33 to 140 nm and temperatures of 300, 500 and 700 Kelvin on the thermal conductivity of pristine models were investigated. Studying the thermal behavior of the materials at higher temperatures helped us to better understand the effect of phonon scattering. In the following sections, the effect of the aforementioned parameters on thermal conductance for six different grain boundaries will be assessed.

In past numerical studies, it was observed that the optimized Tersoff potential, proposed by Lindsay and Broido [27], accurately predicts the thermal conductivity of pristine h-BN and Graphene [13–20]. Hence, this study employs the Tersoff potential for modeling the atomic interaction, which takes the form [28–31]:

$$U = \frac{1}{2} \sum_{i \neq j} f_c\left(r_{ij}\right) \cdot \left[f_R\left(r_{ij}\right) + b_{ij} f_A\left(r_{ij}\right)\right] \tag{1}$$

f_c being a cutoff function, r_{ij} is the distance between atom i and j, b_{ij} accounts for the bond order and the local environment, f_R and f_A are the repulsive and attractive pair potentials, respectively.

Figure 1 shows the pristine molecular dynamic (MD) model for extracting the thermal conductivity of single-layer h-BN and Graphene films with the length (denoted as *L*) of 33 nm. Similar to the work in [20], periodic boundary conditions were employed along the width of the h-BN and graphene sheets in their planar directions for all the developed MD models. Hence, the results obtained implicate an infinite system of grain boundaries.

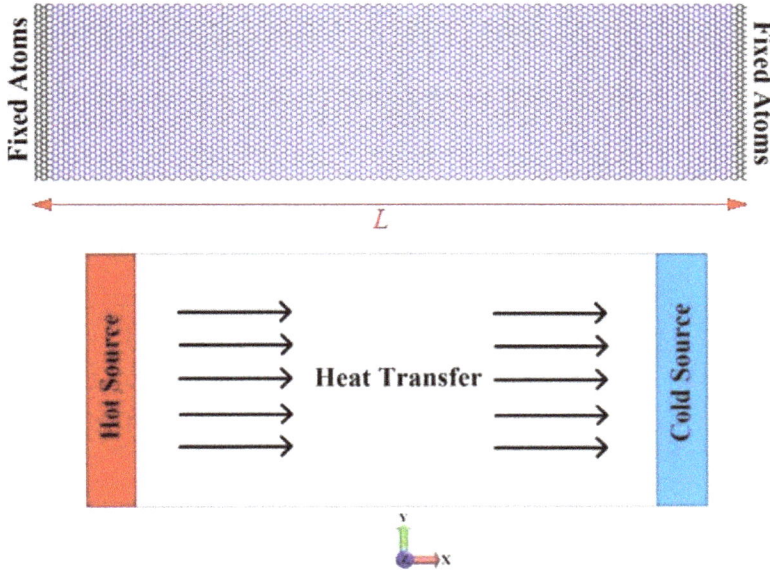

Figure 1. The schematic model for the evaluation of the thermal conductivity of single-layer h-BN and Graphene films with L (denotes the length) ranging 33 to 140 nm.

Figure 2 shows the MD model developed for computing the thermal conductance for both materials, which is composed of two pristine sheets (red and blue sheets) that are connected by six types of grain boundary (as shown in Figure 3). We then divided the sample into 20 slabs excluding the fixed zones at both ends. In Figure 2, L_L and L_R are the length of the left and right sheets, respectively.

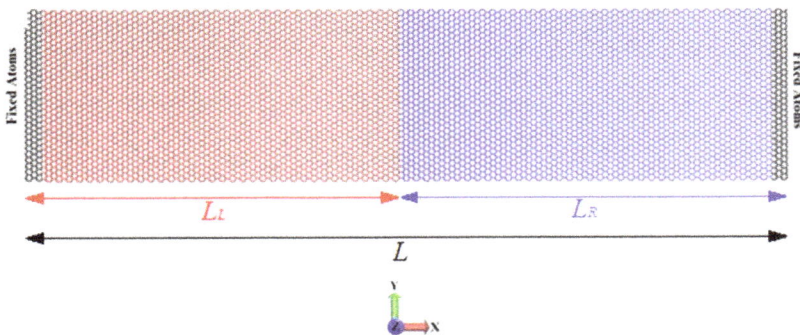

Figure 2. The developed molecular model for the investigation of the thermal conductance of the h-BN and Graphene sheets.

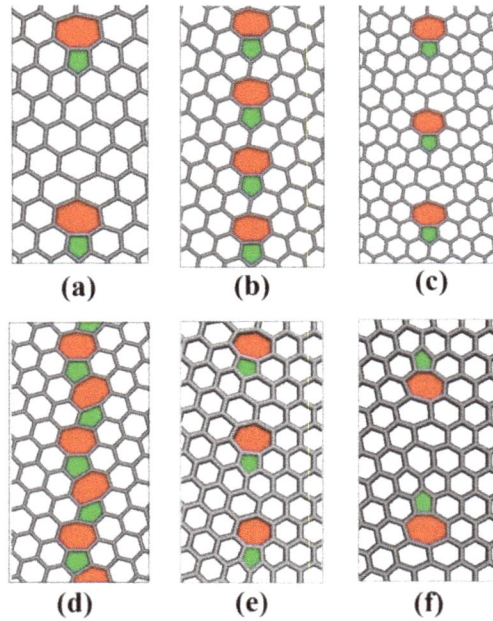

Figure 3. The applied grain boundaries between two separated h-BN and Graphene layers. (**a**) Symmetric GB's with vectors (4, 3) | (4, 3) at armchair oriented tilt angle 9.43° and formation energy 2.83 eV/nm, containing 5 hexagonally bonded atoms between each defect pair; (**b**) Symmetric GB's with vectors (2, 1) | (2, 1) at armchair oriented tilt angle 21.8° and formation energy 3.44 eV/nm, containing 1 hexagonally bonded atoms between each defect pair; (**c**) Symmetric GB's with vectors (3, 2) | (3, 2) at armchair oriented tilt angle 13.2° and formation energy -i1 3.51 eV/nm, containing 3 hexagonally bonded atoms between each defect pair; (**d**) Symmetric GB's with vectors (3, 1) | (3, 1) at armchair oriented tilt angle 32.2° and formation energy -i2 2.91 eV/nm, containing 0 hexagonally bonded atoms between each defect pair; (**e**) Non-Symmetric GB's with vectors (3, 1) | (2, 2) at zigzag oriented tilt angles 13.89° & 30° and formation energy -i1 4.82 eV/nm, containing 2 hexagonally bonded atoms between each defect pair; (**f**) Non-Symmetric GB's with vectors (4, 2) | (3, 3) at zigzag oriented tilt angles 19.1° & 30° and formation energy 3.23 eV/nm, containing 4 hexagonally bonded atoms between each defect pair.

To study the thermal conductance between two separate h-BN and Graphene sheets in Figure 2, we constructed six grain boundaries shown in Figure 3. The 6 plots represent the defect patterns with varying gap sizes filled with hexagonal boron-nitrogen or graphene-graphene bonds. The rate of phonon scattering along the grain boundaries has a direct correlation with the increase in defect concentration. Hence, from several possible combinations of heptagon-pentagon shaped defect pairs, we have chosen the grain boundaries in which the defect concentration ranges from nil to a few dislocations located in-between the defect pairs. Furthermore, our choice of grain boundary selection allows us to study the change in thermal resistivity as the distance between the defect pairs gradually changes. In the following section, the procedure to construct the grain boundaries is described:

The grain boundaries as depicted in Figure 3 are constructed based on the number of hexagonal lattices presented between each heptagon-pentagon (5–7 pair) pair along the line of defect.

As explained in [32], the complete polycrystalline h-BN lattice sheets are a combination of two individual pristine h-BN lattice sheets joined together. The common side of the sheet forms the molecular grain boundaries based on the angle at which one sheet or both sheets are tilted.

The misorientation angle θ is the sum of both angles of the left and right domains and can be computed by $\theta = \theta_L + \theta_R$, where θ_L and θ_R are the corresponding angles in the left or right domain of the polycrystalline h-BN sheets in their respective crystallographic orientations. Each domain is characterized by the translation vectors, n_L, m_L, n_R and m_R. These vectors represent the left and right domains of the lattice structure and are represented as $(n_L, m_L) \mid (n_R, m_R)$ [22,32].

According to [33], the translation vectors are responsible for the periodic arrangement of the defects along the grain boundaries. The following expression explains the technique used to calculate the angle of misorientation by using the corresponding translation vectors.

$$\theta = \theta_L + \theta_R = tan^{-1}\left(\frac{\sqrt{3m_L}}{m_L + 2n_L}\right) + tan^{-1}\left(\frac{\sqrt{3m_R}}{m_R + 2n_R}\right) \tag{2}$$

The misorientation angle of the armchair oriented GB's can be determined from the misorientation angle of the zigzag oriented GB's, through the relation:

$$\theta_{armchair} = 60^{\circ} - \theta_{zigzag} \tag{3}$$

From the values of the translational vectors representing the left and right domains and using them in Equation (2), the misorientation angles were calculated and the symmetric grain boundaries (Figure 3a–d) could be constructed. GB.1–4 represent the symmetric GB's.

- GB.1—(4, 3) ∣ (4, 3) 2.83 eV/nm = 9.43°
- GB.2—(2, 1) ∣ (2, 1) 3.44 eV/nm = 21.78°
- GB.3—(3, 2) ∣ (3, 2) -i1 3.51 eV/nm = 13.17°
- GB.4—(3, 1) ∣ (3, 1) -i2 2.91 eV/nm = 32.2°

Since the angles calculated for the symmetric grain boundaries are oriented in a zigzag manner, they must be converted to their armchair orientation using Equation (3).

The non-symmetric grain boundaries (Figure 3e,f) were also constructed using Equation (2), but were not converted into their armchair orientation due to the irregularity in the positioning of the heptagon-pentagon pairs along the line of defect. From their translation vectors, the misorientation angles for the non-symmetric grain boundaries were calculated. GB.5 and 6 represent the non-symmetric GB's.

- GB.5—(3, 1) ∣ (2, 2) 3.23 eV/nm = 13.89° & 30°
- GB.6—(4, 2) ∣ (3, 3) 3.23 eV/nm = 19.1° & 30°

During the construction of the chosen non-symmetric grain boundaries, the formation of adatoms (additional atoms in the heptagon ring) was unavoidable and hence, a single carbon or boron atom had to be implemented in the molecular structure to overcome this issue. Edge-dislocations were particularly observed during the construction of non-symmetric grain boundaries and by reorganizing the atomic structure it was possible to successfully construct them at the desired misorientation angle. Strong out-of-plane corrugations may result from the presence of dislocations on the structure. This has been further investigated in [34]. By choosing the above-mentioned misorientation angles the gap between neighboring defects were smaller, which facilitates our calculations. More details can be found in [22].

In this study, we employed the non-equilibrium molecular dynamics (NEMD) method [35]. In order to avoid atomic discretion, the atoms are fixed at the beginning and at the end of the model. (Depicted as the black atoms in Figures 1 and 2). Then, we divided the atomic sheet into several strips and imposed heat conduction to atomic sheets. The temperature at each strip may be determined as:

$$T_i(Strip) = \frac{2}{3N_i k_B}\sum_i \frac{p_i^2}{2m_i} \tag{4}$$

where, T_i denotes the temperature of ith strip, N_i is the number of atoms in ith strip, k_B is the Boltzmann's constant, m_j is atomic mass of atom jth and P_j indicates the atomic momentum of atom j, respectively.

h-BN and Graphene sheets were firstly relaxed at 300 K using the Nosé–Hoover thermostat (NVT) method [36] for 50 ps. Then, the first and last strips were subjected to a temperature difference of 20 K using the NVT method, while the other strips were governed at a desired temperature and simulated at constant energy level using the NVT method and NVE ensemble, respectively. Hot and cold sources were assigned between the first and twentieth slabs, respectively (Figures 1 and 2). After 200 ps the system reached (heat transfer) a steady-state condition including a transient temperature difference. After this stage the heat conduction was imposed to the models for 4800 ps, while the energy of the atoms at the hot strip increases slightly and simultaneously, the energy of the atoms at the cold strip decreases slightly. This way, the thermal properties can be computed using the average of temperature at each strip.

3. Obtained Results for Thermal Conductivity

To obtain more accurate results from the NEMD simulations, the total energy of the system should remain constant. In other words, the quantity of the increased and decreased energy in the hot and cold sources must be the same. Figure 4 explains the possible fluctuation in energy of the atoms computed at the hot- and cold-strips for both models as well as the pristine sheets. The slope value for both hot-strip and cold-strips are nearly identical by means of adding and removing energy from the hot and cold reservoirs constantly, the energy of the atoms is constant at the conducting strips. Figure 4 depicts the evolution of the energy added to the hot reservoir and removed from the cold reservoir per unit length of the h-BN and graphene sheets. The curves are following a linear pattern, which demonstrates that the total energy of the system remains constant during the analysis that in turn is caused by applying a constant heat flux given by

$$q_x = \frac{dq}{A \cdot dt} \tag{5}$$

where, q is the added or removed energy, A is the cross sectional area of the sheet and dt is the time increment. Furthermore, the temperature gradient can be calculated based on the slope of the dotted red line in Figure 5. It is worth noting that, for the sake of better accuracy of dT/dx, the results of the average temperature of the first and last three layers (as shown in Figure 5) were ignored.

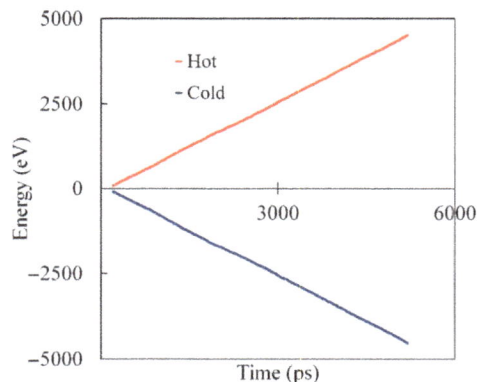

Figure 4. The energy-time graph (the energy added to the hot reservoir and removed from the cold reservoir) for a sheet with the length of 33 nm. The heat flux in Equation (5) can be calculated based on the slope of each line. The slope of two drawn linear curves must be the same because the energy added in the hot source is equal to the energy removed in the cold source during the simulations.

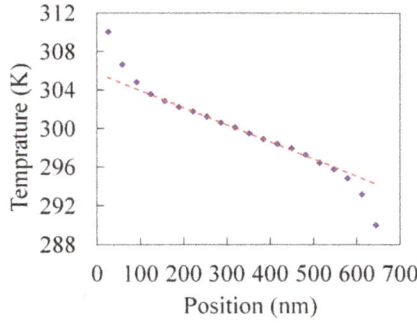

Figure 5. The temperature gradient along the longitudinal direction including the average temperature of all twenty layers for a single-layer sheet with length of 33 nm at 300 Kelvin. To calculate a linear relation of dT/dx, the temperature gradient without the average temperature of the first and last three layers was considered and dT/dx was determined according to the slope of the dotted red line.

Based on the calculated heat fluxes (q_x) and the temperature gradient ($\frac{dT}{dx}$), the thermal conductivity (k) of h-BN, and Graphene sheets was calculated using the 1D form of Fourier law as follows:

$$k = \frac{q_x}{A_{cross}\frac{dT}{dx}} \tag{6}$$

where q_x is the heat flux, A_{cross} is cross sectional area of the sheet, and $\frac{dT}{dx}$ is the temperature gradient. In Figure 5, it is assumed that the temperature remains constant at both ends of the lattice sheet. Hence, in a pristine lattice sheet, the temperature tends to drop gradually as the heat flows from the hotter end to the colder end. This assumption is confirmed by the correlation values obtained from the MD simulations, showing a linear flow of heat across the pristine lattice sheets.

We carried out MD simulations for models of lengths ranging from 33 to 140 nm. Figure 6 shows the inverse of computed thermal conductivity for h-BN and graphene models as a function of inverse length. The thermal conductivity pristine models with infinite length was calculated by fitting a linear relation. It can be clearly seen in Figure 6 that the thermal conductivity increases with increasing length of the sheets. In contrast, Figure 6 demonstrates that the thermal conductivity of the sheets decreases at higher temperature due to the increase in phonon-phonon scattering. As the temperature increases, the phonon lifetime drops and this leads to the convergence of the thermal conductivity at lower correlation times [13].

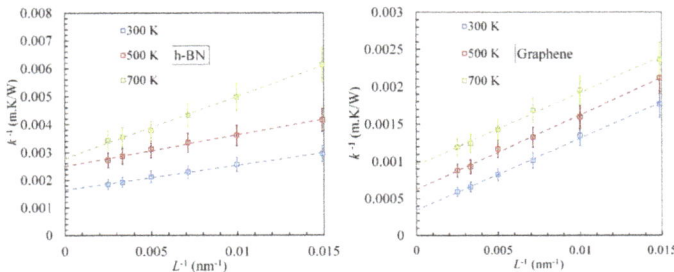

Figure 6. The calculated inverse of thermal conductivity vs. inverse of length of h-BN and Graphene sheets based on the variation of sheet length and temperature. To extrapolate the thermal conductivity for infinite sheets, the inverse of the thermal conductivity is plotted versus the inverse of the length. The dashed-line is employed to demonstrate the length dependence of the thermal conductivity.

Table 1 represents the computed thermal conductivity of h-BN and Graphene models for 3 temperatures. It should be noted that the thermal conductivity of infinitely long sheets can be computed by extrapolating the results for finite lengths. Hence, the term effective phonon mean free path (Λ_{eff}) can be explained as:

$$\frac{1}{\Lambda_{eff}} = \frac{1}{\Lambda} + \frac{1}{L} \tag{7}$$

where, L is the length of the pristine sheet. With attention to proportionality of thermal conductivity to Λ_{eff}, the term thermal conductivity for an infinite sheet can be computed by extrapolating the values to $1/L = 0$ [37–39]. Consequently, the effective thermal conductivity can be determined by the one-dimensional form of the Fourier law:

$$k_{eff} = q\frac{L}{\Delta T} \tag{8}$$

where, L denotes the length of the cross-section, q is the heat flux of the system, and ΔT is the temperature difference along the laminate. The predicted thermal conductivity for a single-layer h-BN is 606.6 W/m· K, which is in good agreement with the predictions of 600 W/m·K in [13], and the measured experimental value of 390 W/m·K for in-plane thermal conductivity of bulk h-BN [14]. We obtained a thermal conductivity of single-layer Graphene of 2941.2 W/m·K, which agrees well with the reported experimental data ranging from 3000 to 5000 W/m· K at room temperature (300 K) [40–42]. Our predictions also agree well with other predictions in the literature; see e.g., the contribution in [38] reporting a value for the thermal conductivity of graphene around 2650 W/m·K. In addition, Hahn et al. predicted the thermal conductivity of 262 W/m·K for a single-crystalline graphene with the sheet length of 200 nm using approach-to-equilibrium molecular dynamics (AEMD) simulation [43].

Table 1. The obtained temperature dependent results of thermal conductivity in h-BN and Graphene models.

	Thermal Conductivity (W/m·K)		
	300 Kelvin	**500 Kelvin**	**700 Kelvin**
h-BN	606.58	400.26	359.54
Graphene	2941.18	1596.17	1048.22

4. Obtained Results for Thermal Conductance

The thermal conductance of h-BN and Graphene sheets along the six modeled grain boundaries depicted in Figure 3 was calculated based on the variation of the sheet length and temperature. The thermal conductance for each model takes the form:

$$C = \frac{q_x}{A_{cross}\cdot\Delta T} \tag{9}$$

where, q_x is the heat flux, A_{cross} is the cross sectional area of the sheet and ΔT is the temperature difference along the interface of two sheets. The heat flux can be calculated using the slope of the energy curves as shown in Figure 7. As explained in the previous section, the slope of two curves must be close to each other since the total amount of energy must be constant during the analysis.

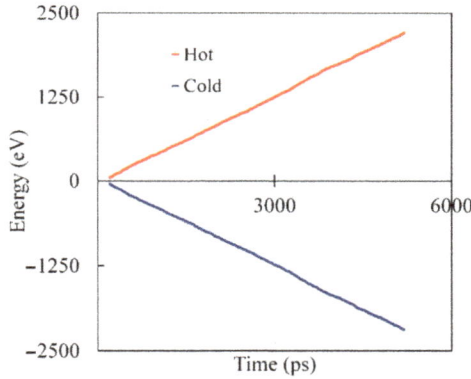

Figure 7. The energy-time graph for a sheet with the length of 33 nm. The heat flux in Equation (5) can be calculated based on the slope of each line. The slope of two drawn linear curves must be the same because the energy added in the hot source is equal to the energy removed in cold source during the simulations.

Figure 8 illustrates the temperature gradient for 20-layer at room temperature. For the sake of conciseness, only the temperature gradient for a sheet with a length of 33 nm is presented. The temperature difference (ΔT) at the grain boundaries can be computed based on the temperature-position graph. A sudden drop in the temperature is seen along the transition regions of the two domains, which is denoted by the gap in the plotted area. The gap indicates the presence of the grain boundaries and confirms the non-linearity in the heat flow across a polycrystalline sheet. After the heat is transferred across the line of defects, the flow of heat is linear again. This is confirmed by their respective slopes, which shows a linear correlation.

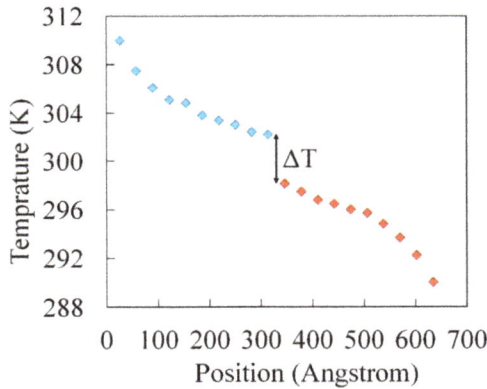

Figure 8. Temperature difference at the grain boundary for a single-layer sheet with length of 33 nm at 300 Kelvin. This temperature difference should be used in Equation (5) for the calculation of the thermal conductance.

Figure 9 exhibits the computed thermal conductance for h-BN models for six grain boundaries at different temperatures and lengths. It can be seen in Figure 9 that that the term of thermal contact conductance for h-BN material is almost independent of temperature and length [13]. As can be seen from Figure 9, the thermal conductance for h-BN for different sheet lengths, temperatures, and grain boundaries varies between 11.1 to 30.3 GW/m²·K, which agrees well with the reported values of

$11 \pm 1 \, \text{GW/m}^2 \cdot \text{K}$ in [13]. The insignificant influence of the temperature on the thermal conductance at grain boundaries can be explained as the heat transfer along the grain boundaries is dominated by the phonon-defect scattering. As expected, Figure 9 demonstrates that the variation of the sheet length does not influence the thermal conductance of h-BN films significantly. The maximum variation of thermal conductance for different sheet lengths with identical grain boundary and temperature is 16%.

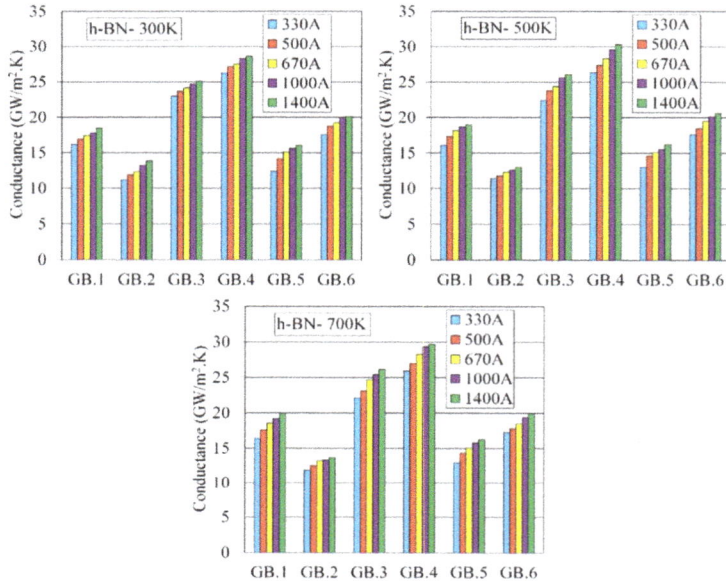

Figure 9. Predicted the term of thermal conductance of h-BN material for six grain boundaries, sheet lengths (330 A, 500 A, 670 A, 1000 A and 1400 A) and three temperatures (300 K, 500 K and 700 K).

Figure 10 depicts the thermal conductance of graphene different grain boundaries in terms of different sheet lengths and temperatures. It can be seen in Figure 10 that the predicted thermal conductance of Graphene varies between 34.3 to 68.1 $\text{GW/m}^2 \cdot \text{K}$. The experimentally measured thermal conductance of graphene was reported around 133 $\text{GW/m}^2 \cdot \text{K}$ [4], while the predictions vary from 15 to 45 $\text{GW/m}^2 \cdot \text{K}$ [44] and from 0.14 to 0.58 $\text{GW/m}^2 \cdot \text{K}$ [43] using the NEMD and AEMD approaches, respectively. In the case of identical grain boundary and temperature, the maximum variation thermal conductance of graphene material for five different sheet lengths is around 8%.

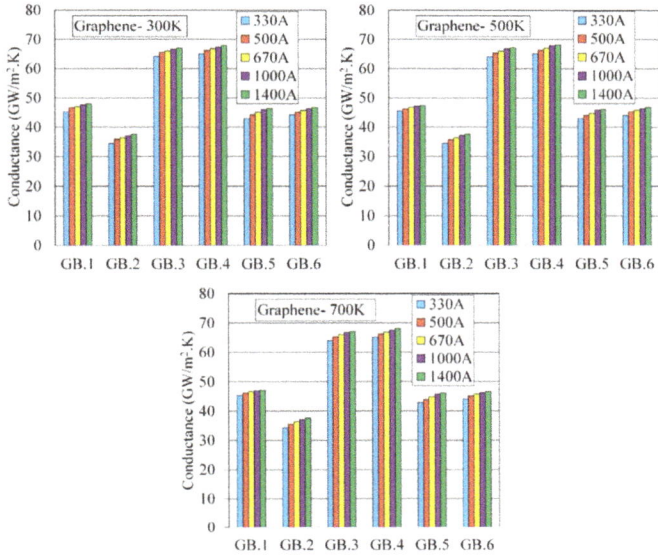

Figure 10. Predicted the term of thermal conductance of Graphene material for six grain boundaries, sheet lengths (330 A, 500 A, 670 A, 1000 A and 1400 A) and three temperatures (300 K, 500 K and 700 K).

For the sake of better understanding, the variation of thermal conductance for the six studied grain boundaries in terms of temperature is illustrated in Figure 11. In Figure 11, the term of thermal conductance was calculated using the same methodology for computing the thermal conductivity (cf. Equation (7)). In Figure 11a,b, the thermal conductance for six employed grain boundaries varies between 14.3 to 31.5 GW/m^2·K and 38.5 to 69.2 GW/m^2·K for h-BN and Graphene, respectively.

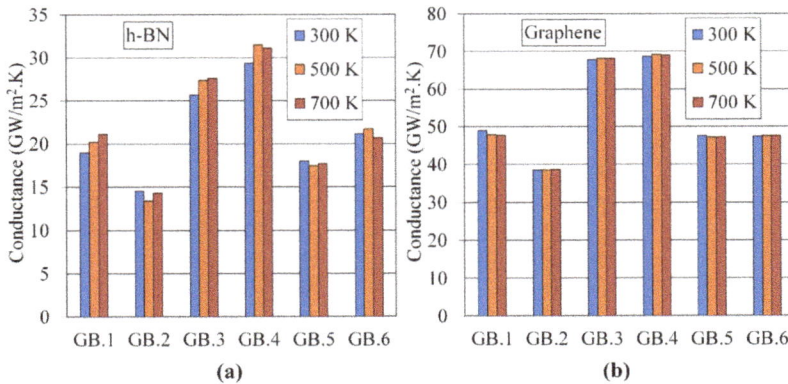

Figure 11. Computed the term of thermal conductance of graphene material for six employed grain boundaries at three temperatures (300 K, 500 K and 700 K) and six grain boundaries. (**a**) h-BN; (**b**) Graphene.

Figure 11 demonstrates that the thermal conductance of h-BN and graphene is independent of the temperature and sheet length. However, small discrepancies in the predicted thermal conductivities for h-BN films can be observed in Figure 11a.

5. Summary and Conclusions

In this study, we performed numerical simulations to study the influence of temperature and sheet length on thermal properties (i.e., thermal conductivity and thermal conductance) of two materials; i.e., hexagonal Boron Nitride (h-BN) and graphene. Six grain boundaries with different defect concentrations were used to investigate the thermal conductance along hBN-hBN and Graphene-Graphene interfaces. The non-equilibrium molecular dynamics method (NEMD) was employed to simulate and analyze the thermal properties. To verify our results, we first calculated the thermal conductivity of pristine h-BN and Graphene models and compared these values with the previous results from the literatures. The computed thermal conductivities of h-BN and Graphene sheets were close to the measured thermal conductivities of pristine h-BN and Graphene. The results show that by increasing the sheet lengths, the thermal conductivity in h-BN and Graphene models increases and by increasing the temperature, the thermal conductivity decreases. On the other hand, the thermal conductance of the material is independent of the sheet length and temperature.

Author Contributions: Conceptualization, M.R.A.K. and R.P.U.; Methodology, T.R.; Software, M.R.A.K. and R.P.U.; Validation, M.R.A.K. and T.R.; Formal Analysis, M.R.A.K.; Investigation, M.M. and A.H.N.S.; Resources, T.R.; Data Curation, M.M. and A.H.N.S.; Writing-Original Draft Preparation, M.R.A.K., R.P.U. and T.R.; Writing-Review & Editing, T.R., A.H.N.S., M.M.; Visualization, M.R.A.K. and R.P.U.; Supervision, T.R.; Project Administration, T.R.; Funding Acquisition, T.R.

Conflicts of Interest: The authors declare that there is no conflict of interest regarding the publication of this paper.

References

1. Novoselov, K.S.; Geim, A.K.; Morozov, S.V.; Jiang, D.A.; Zhang, Y.; Dubonos, S.V.; Grigorieva, I.V.; Firsov, A.A. Electric field effect in atomically thin carbon films. *Science* **2004**, *306*, 666–669. [CrossRef] [PubMed]

2. Novoselov, K.S.; Jiang, D.; Schedin, F.; Booth, T.J.; Khotkevich, V.V.; Morozov, S.V.; Geim, A.K. Two-dimensional atomic crystals. *Proc. Natl. Acad. Sci. USA* **2005**, *102*, 10451–10453. [CrossRef] [PubMed]

3. Geim, A.K.; Novoselov, K.S. The rise of graphene. *Nat. Mater.* **2007**, *6*, 183–191. [CrossRef] [PubMed]

4. Ghosh, S.; Bao, W.; Nika, D.L.; Subrina, S.; Pokatilov, E.P.; Lau, C.N.; Balandin, A.A. Dimensional crossover of thermal transport in few-layer graphene. *Nat. Mater.* **2010**, *9*, 555–558. [CrossRef] [PubMed]

5. Xu, X.; Pereira, L.F.; Wang, Y.; Wu, J.; Zhang, K.; Zhao, X.; Bae, S.; Bui, C.T.; Xie, R.; Thong, J.T.L.; et al. Length-dependent thermal conductivity in suspended single-layer graphene. *Nat. Commun.* **2014**, *5*, 3689. [CrossRef] [PubMed]

6. Watanabe, K.; Taniguchi, T.; Kanda, H. Direct-bandgap properties and evidence for ultraviolet lasing of hexagonal boron nitride single crystal. *Nat. Mater.* **2004**, *3*, 404–409. [CrossRef] [PubMed]

7. Jiao, X.; Young, J.S.; Jin, H.P.; Sungjoo, L. Graphene/black phosphorus heterostructured photodetector. *Solid-State Electron.* **2018**, *144*, 86–89.

8. Hosseinzadeh, A.; Bidmeshkipour, S.; Abdi, Y.; Arzi, E.; Mohajerzadeh, S. Graphene based strain sensors: A comparative study on graphene and its derivatives. *Appl. Surf. Sci.* **2018**, *448*, 71–77. [CrossRef]

9. Le, Z.; Guangmei, H.; Wei, Z.; Qing, A.; Jinkui, F.; Lin, Z.; Pengchao, S.; Lijie, C. Alumium/graphene composites with enhanced heat-dissipation properties by in-situ reduction of graphene oxide on aluminum particles. *J. Alloys Compd.* **2018**, *748*, 854–860.

10. Shirazi, A.H.N.; Mohebbi, F.; Azadi Kakavand, M.R.; He, B.; Rabczuk, T. Paraffin Nanocomposites for Heat Management of Lithium-Ion Batteries. *J. Nanomater.* **2016**, *59*. [CrossRef]

11. Giannopoulos, G.I. On the buckling of hexagonal boron nitride nanoribbons via structural mechanics. *Superlattices Microstruct.* **2018**, *115*, 1–9. [CrossRef]

12. Tongwei, H.; Scarpa, F.; Allan, N.L. Super stretchable hexagonal boron bitride Kirigami. *Thin Solid Films* **2017**, *632*, 35–43.

13. Mortazavi, B.; Pereira, L.F.C.; Jiang, J.W.; Rabczuk, T. Modelling heat conduction in polycrystalline hexagonal boron-nitride thin films. *Sci. Rep.* **2015**, *5*, 13228. [CrossRef] [PubMed]

14. Sichel, E.; Miller, R.; Abrahams, M.; Buiocchi, C. Heat capacity and thermal conductivity of hexagonal pyrolytic boron nitride. *Phys. Rev. B* **1976**, *13*, 4607–4611. [CrossRef]

15. Zhang, Y.B.; Tan, Y.W.; Stormer, H.L.; Kim, P. Experimental observation of quantum hall effect and Berry's phase in graphene Nature. *Nat. J. Sci.* **2005**, *438*, 201–204. [CrossRef] [PubMed]

16. Lee, C.; Wei, X.; Kysar, J.W.; Hone, J. Measurement of the elastic properties and intrinsic strength of monolayer graphene. *Science* **2008**, *321*, 385–388. [CrossRef] [PubMed]

17. Williams, J.R.; DiCarlo, L.; Marcus, C.M. Quantum hall effect in a gate-controlled pn junction of graphene. *Science* **2007**, *317*, 638–641. [CrossRef] [PubMed]

18. Stankovich, S.; Dikin, D.A.; Dommett, G.H.; Kohlhaas, K.M.; Zimney, E.J.; Stach, E.A.; Piner, R.D.; Nguyen, S.T.; Ruoff, R.S. Graphene-based composite materials Nature. *Nat. J. Sci.* **2006**, *442*, 282–286. [CrossRef] [PubMed]

19. Mortazavi, B.; Cuniberti, G. Atomistic modeling of mechanical properties of polycrystalline graphene. *J. Nanotechnol.* **2014**, *25*, 215704. [CrossRef] [PubMed]

20. Mortazavi, B.; Pötschke, M.; Cuniberti, G. Multiscale modeling of thermal conductivity of polycrystalline graphene sheets. *J. Nanoscale* **2014**, *6*, 3344–3352. [CrossRef] [PubMed]

21. Han, J.; Ryu, S.; Sohn, D.; Im, S. Mechanical strength characteristics of asymmetric tilt grain boundaries in graphene. *J. Carbon* **2014**, *68*, 250–257. [CrossRef]

22. Zhang, J.; Zhao, J. Structures and electronic properties of symmetric and nonsymmetric graphene grain boundaries. *J. Carbon* **2013**, *55*, 151–159. [CrossRef]

23. Plimpton, S. Fast parallel algorithms for short-range molecular dynamics. *J. Comput. Phys.* **1995**, *117*, 1–19. [CrossRef]

24. Liu, Z.; Gong, Y.; Zhou, W.; Ma, L.; Yu, J.; Idrobo, J.C.; Jung, J.; MacDonald, A.H.; Vajtai, R.; Lou, J.; et al. Ultrathin high-temperature oxidation-resistant coatings of hexagonal boron nitride. *Nat. Commun.* **2013**, *4*, 2541. [CrossRef] [PubMed]

25. Hou, X.; Yu, Z.; Li, Y.; Chou, K. Preparation and properties of hexagonal boron nitride fibers used as high temperature membrane filter. *Mater. Res. Bull.* **2014**, *49*, 39–43. [CrossRef]

26. Veronese, G.P.; Allegrezza, M.; Canino, M.; Centurioni, E; Ortolani, L.; Rizzoli, R.; Morandi, V.; Summonte, C. Graphene as transparent conducting layer for high temperature thin film device applications. *Sol. Energy Mater. Sol. Cells* **2015**, *138*, 35–40. [CrossRef]

27. Lindsay, L.; Broido, D.A. Optimized Tersoff and Brenner empirical potential parameters for lattice dynamics and phonon thermal transport in carbon nanotubes and graphene. *Phys. Rev. B Condens. Matter.* **2010**, *82*, 205441. [CrossRef]

28. Tersoff, J. Empirical interatomic potential for carbon, with applications to amorphous carbon. *Phys. Rev. Lett.* **1988**, *61*, 2879. [CrossRef] [PubMed]

29. Tersoff, J. New empirical approach for the structure and energy of covalent systems. *Phys. Rev. B* **1988**, *37*, 6991–7000. [CrossRef]

30. Gillespie, B.A. *Bond Order Potentials for Group IV Semiconductors*; University of Virginia: Charlottesville, VA, USA, 2009.

31. Tersoff, J. Emprical interatomic potential for silicon with improved elastic properties. *Phys. Rev. B* **1988**, *38*, 9902–9905. [CrossRef]

32. Liu, T.; Gajewski, G.; Pao, C.; Chang, C. Structure, energy and structural transformations of graphene grain boundaries from atomistic simulations. *Carbon* **2011**, *49*, 2306–2317. [CrossRef]

33. Minkoff, I.; Myron, S. Rotation boundaries and crystal growth in the hexagonal system. *Philos. Mag.* **1969**, *19*, 379–387. [CrossRef]

34. Lehtinen, O.; Kurasch, S.; Krasheninnikov, A.V.; Kaiser, U. Atomic scale study of life cycle of a dislocation in graphene from birth to annihilation. *Nat. Commun.* **2013**, *4*, 2098. [CrossRef] [PubMed]

35. Ciccotti, G.; Kapral, R.; Sergi, A. *Non-Equilibrium Molecular Dynamics, Handbook of Materials Modeling: Methods*; Springer: Berlin, Germany, 2005; pp. 745–761.

36. Evans, D.J.; Holian, B.L. The Nose–Hoover thermostat. *J. Chem. Phys.* **1985**, *83*, 4069–4074. [CrossRef]

37. Schelling, P.K.; Phillpot, S.R.; Keblinski, P. Comparison of atomic-level simulation methods for computing thermal conductivity. *Phys. Rev. B.* **2002**, *65*, 144306. [CrossRef]

38. Mortazavi, B.; Rabczuk, T. Multiscale modeling of heat conduction in graphene laminates. *Carbon* **2015**, *85*, 1–7. [CrossRef]

39. He, Y.; Savić, I.; Donadio, D.; Galli, G. Lattice thermal conductivity of semiconducting bulk materials: Atomistic simulations. *Phys. Chem. Chem. Phys.* **2012**, *14*, 16209–16222. [CrossRef] [PubMed]

40. Balandin, A.A.; Ghosh, S.; Bao, W.; Calizo, I.; Teweldebrhan, D.; Miao, F.; Lau, C.N. Superior thermal conductivity of single-layer graphene. *Nano Lett.* **2008**, *8*, 902–907. [CrossRef] [PubMed]

41. Ghosh, S.; Calizo, I.; Teweldebrhan, D.; Pokatilov, E.P.; Nika, D.L.; Balandin, A.A.; Bao, W.; Miao, F.; Lau, C.N. Extremely high thermal conductivity of graphene: Prospects forthermal management applications in nanoelectronic circuits. *Appl. Phys. Lett.* **2008**, *92*, 151911. [CrossRef]

42. Narasaki, M.; Wang, H.; Nishiyama, T.; Ikuta, T.; Takahashi, K. Experimental study on thermal conductivity of free-standing fluorinated single-layer graphene. *Appl. Phys. Lett.* **2017**, *111*, 093103. [CrossRef]

43. Hahn, K.R.; Melis, C.; Colombo, L. *Thermal Transport in Nanocrystalline Graphene: The Role of Grain Boundaries*; InGraphITA, Springer: Cham, Switzerland, 2017; pp. 1–17.

44. Bagri, A.; Kim, S.P.; Ruoff, R.S.; Shenoy, V.B. Thermal transport across twin grain boundaries in polycrystalline graphene from nonequilibrium molecular dynamics simulations. *Nano Lett.* **2011**, *11*, 3917. [CrossRef] [PubMed]

energies

MDPI

Article

Boron Monochalcogenides; Stable and Strong Two-Dimensional Wide Band-Gap Semiconductors

Bohayra Mortazavi [1,2,*] and Timon Rabczuk [3]

1   Institute of Structural Mechanics, Bauhaus-Universität Weimar, Marienstr. 15, D-99423 Weimar, Germany
2   Institute for Materials Science and Max Bergman Center of Biomaterials, TU Dresden,
     01062 Dresden, Germany
3   College of Civil Engineering, Department of Geotechnical Engineering, Tongji University, Shanghai 200092,
     China; timon.rabczuk@uni-weimar.de
*   Correspondence: bohayra.mortazavi@gmail.com; Tel.: +49-157-8037-8770; Fax: +49-364-358-4511

Received: 13 May 2018; Accepted: 13 June 2018; Published: 15 June 2018

Abstract: In this short communication, we conducted first-principles calculations to explore the stability of boron monochalcogenides (BX, X = S, Se or Te), as a new class of two-dimensional (2D) materials. We predicted BX monolayers with two different atomic stacking sequences of ABBA and ABBC, referred in this work to 2H and 1T, respectively. Analysis of phonon dispersions confirm the dynamical stability of BX nanosheets with both 2H and 1T atomic lattices. Ab initio molecular dynamics simulations reveal the outstanding thermal stability of all predicted monolayers at high temperatures over 1500 K. BX structures were found to exhibit high elastic modulus and tensile strengths. It was found that BS and BTe nanosheets can show high stretchability, comparable to that of graphene. It was found that all predicted monolayers exhibit semiconducting electronic character, in which 2H structures present lower band gaps as compared with 1T lattices. The band-gap values were found to decrease from BS to BTe. According to the HSE06 results, 1T-BS and 2H-BTe show, respectively, the maximum (4.0 eV) and minimum (2.06 eV) electronic band gaps. This investigation introduces boron monochalcogenides as a class of 2D semiconductors with remarkable thermal, dynamical, and mechanical stability.

Keywords: two-dimensional semiconductor; first-principles; mechanical; thermal

1. Introduction

After the great success of graphene [1–3], during the last decade the interest in two-dimensional (2D) materials has been continuously increased. One of the critical factors that has contributed largely in astonishing advances in the field of 2D materials is the available advanced theoretical techniques that enabled researchers to examine the stability of new compositions and estimate properties and suggest possible synthesis routes [4–7] using pure computational experiments. As an example, in 2015, successful experimental realization of borophene nanosheets were reported through epitaxial growth of boron atoms on a Ag surface [8,9]. Nevertheless, prior to these experimental successes, stability, fabrication conditions, and electronic properties of various borophene structures were theoretically predicted [10,11]. Currently, the family of 2D materials includes prominent members, such as: hexagonal boron-nitride [12,13], carbon-nitride 2D networks [14–17], silicene [18,19], germanene [20], transition metal dichalcogenides [21–23], phosphorene [24,25], and indium monochalcogenides [26].

In the most recent theoretical and experimental work by Cherednichenko et al. [27], the phase transition in a rhombohedral boron monosulfide (*r*-BS) bulk layered structure was studied. Interestingly, the atomic structure of *r*-BS in the single-layer form is similar to that of indium selenide, which was recently synthesized in 2D forms by Bandurin et al. [26]. Worthy noting is that the stability of single-layer indium chalcogenides were first confirmed theoretically by Zólyomi and co-workers [28], a few years earlier than the experimental realization [26]. Also noteworthy is that BS monolayers with four different atomic lattices were most recently theoretically predicted by Fan and co-workers [29]. Motivated by the previous investigations [26,28], in this work we decided to examine the stability of single-layer boron monochalcogenides, with a chemical formula of BX, where X stands for S, Se, or Te chalcogen elements. To this aim, we conducted density functional theory calculations to explore the structural and thermal stability, and the basic mechanical and electronic properties of single-layer boron monochalcogenides as a class of 2D materials.

2. Computational Method

First-principles density functional theory (DFT) simulations in this work were conducted using the Vienna ab initio simulation package (VASP) [30–32]. The generalized gradient approximation (GGA) exchange-correlation functional of Perdew-Burke-Ernzerhof (PBE) [33] and plane wave basis set with an energy cut-off of 500 eV were also employed. The conjugate gradient method within the tetrahedron method with Blöchl corrections [34] was used for the geometry optimizations, with termination criteria of 10^{-5} eV and 0.005 eV/Å for the energy and forces, respectively, using a $19 \times 19 \times 1$ Monkhorst-Pack [35] k-point mesh size. Since the PBE functional underestimates the band gap values, we used Heyd-Scuseria-Ernzerhof (HSE06) [36] method to report the electronic band gaps of predicted monolayers. To assess the dynamical stability of BX monolayers, we conducted density functional perturbation theory (DFPT) simulations for 4×4 super-cells within the finite displacement method using the Phonopy package [37]. The thermal stability of single-layer BX was explored via ab initio molecular dynamics (AIMD) simulations for 3×3 super-cells with a time increment of 1 fs and $2 \times 2 \times 1$ k-point mesh size. To evaluate the mechanical properties, we conducted uniaxial tensile modelling for the unit-cells. To this aim, we increased the size of the periodic simulation box along the loading direction with a constant engineering strain step. In order to observe the uniaxial stress condition, the simulation box size along the transverse direction of the loading was changed to reach negligible stress along this direction.

3. Results and Discussions

Like indium chalcogenides [28], predicted BX nanomembranes in this work also show hexagonal lattices, in which two layers of columned B atoms are sandwiched between two chalcogen atom layers. Depending on the arrangement of chalcogen atoms, BX nanosheets can exhibit two different atomic structures. In accordance with transition metal dichalcogenides' atomic lattices [38–40], we call the graphene-like BX with an atomic stacking sequence of ABBA as the 2H structure. On the other hand, BX nanosheets with a 1T phase present an atomic stacking sequence of ABBC, in which the chalcogen atoms, either on the bottom or top, are placed in the hollow center of the hexagonal lattice. Figure 1, illustrates the atomic lattices of 2H- and 1T-BX monolayers. The atomic lattice of BX monolayers can be well defined by the hexagonal lattice constant and B-B and B-X bond lengths. The lattice parameters of energy minimized BX monolayers with hexagonal unit-cells are summarized in Table 1. To compare the stability of 2H and 1T phases, we calculated the energy per atom for BX monolayers and the results are also included in Table 1. Interestingly, boron monochalcogenides with 1T lattice show lower energies and, thus, are more stable in comparison with 2H counterparts.

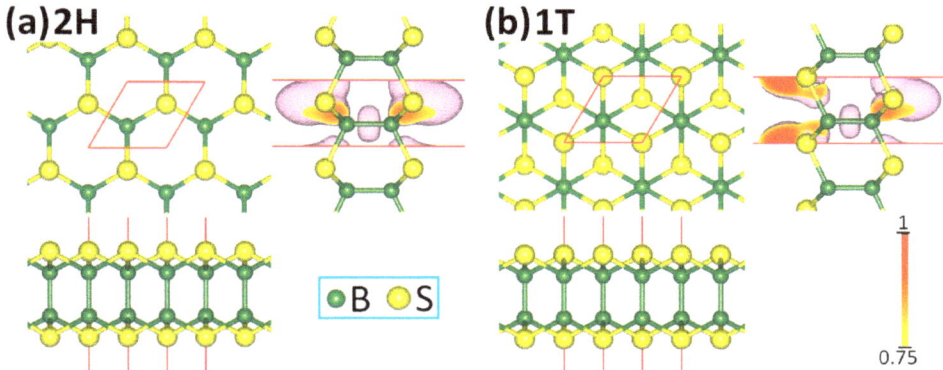

Figure 1. Top and side views of atomic configuration of single-layer BS with 2H and 1T lattices, with ABBA and ABBC stacking sequences, respectively. For the right side views the electron localization function 3D profiles are also plotted. The red box shows the unit-cell which includes four atoms. The VESTA [41] package was employed to illustrate the atomic structures.

Table 1. Lattice parameters of energy minimized BX monolayers. Here, L_{B-B} and L_{B-X}, $E_{unit-cell}$, and ΔQ stand for B-B and B-X bond lengths, energy per atom of a unit-cell, and Bader results for charge transfer from a single B atom to a chalcogen atom, respectively.

Structure		Lattice Constant (Å)	L_{B-B} (Å)	L_{B-X} (Å)	$E_{unit-cell}$ (eV)	ΔQ
BS	2H	3.041	1.728	1.950	−5.813	1.88
	1T	3.056	1.704	1.954	−5.818	1.86
BSe	2H	3.259	1.712	2.087	−5.241	0.68
	1T	3.274	1.686	2.087	−5.256	0.58
BTe	2H	3.565	1.713	2.314	−4.720	−0.1
	1T	3.589	1.681	2.316	−4.746	−0.11

In order to better understand the bonding nature in boron monochalcogenide nanomembranes, we also plotted the electron localization function (ELF) [42] for BS monolayers in Figure 1. It is worth reminding the reader that ELF is a normalized and position-dependent function in which the values close to one corresponds to a high probability of finding electron localizations, and ELF equal to one half corresponds to the region of electron gas-like behaviour. We found that for the all boron monochalcogenides lattices, the ELF values around the centre of all bonds are greater than 0.75, confirming the covalent bonding nature in this class of 2D materials. As is clear in Figure 1, electron localizations also occur around the chalcogen atoms, which is due to their higher valance electrons. To investigate the charge transfer between B and chalcogen atoms, we conducted the Bader charge analysis [43] and the results for different atomic lattices are also included in Table 1. Interestingly, for BS and BSe monolayers, considerably higher electronegativity of S and Se atoms in comparison with B atoms lead to charge gain from B atoms. These charge transfers accordingly induce ionic interactions between heteronuclear bonds in BS and BSe nanomembranes. In these cases, the charge transfer is higher for the 2H phase as compared with 1T lattices, resulting in stiffer bonding and smaller hexagonal lattice constants for 2H lattices. For BTe nanosheets the Bader results, however, suggest that the charge transfer is convincingly negligible.

In order to analyse the dynamical stability of the single-layer BX lattices, the phonon dispersions along the high symmetry Γ-M-K-Γ directions were computed and the obtained results are shown in Figure 2. As is clear for all the predicted monolayers, phonon dispersions are convincingly free of imaginary vibrational frequencies and thus they are all stable from dynamical point of view. For all the structures, in the phonon dispersions two gaps in frequencies are observable. For all the monolayers before the first gap in the phonons frequency, the chalcogen atoms contribute mostly to the vibrations. In contrast, boron atoms with lighter atomic weights were found to dominate the vibrations after the first gap in the phonon dispersions. By increasing the weight of the chalcogen atoms, the gaps in the phonon dispersions widen. For the predicted BX monolayers, 1T lattices were found to exhibit slightly higher frequencies in comparison with 2H phases.

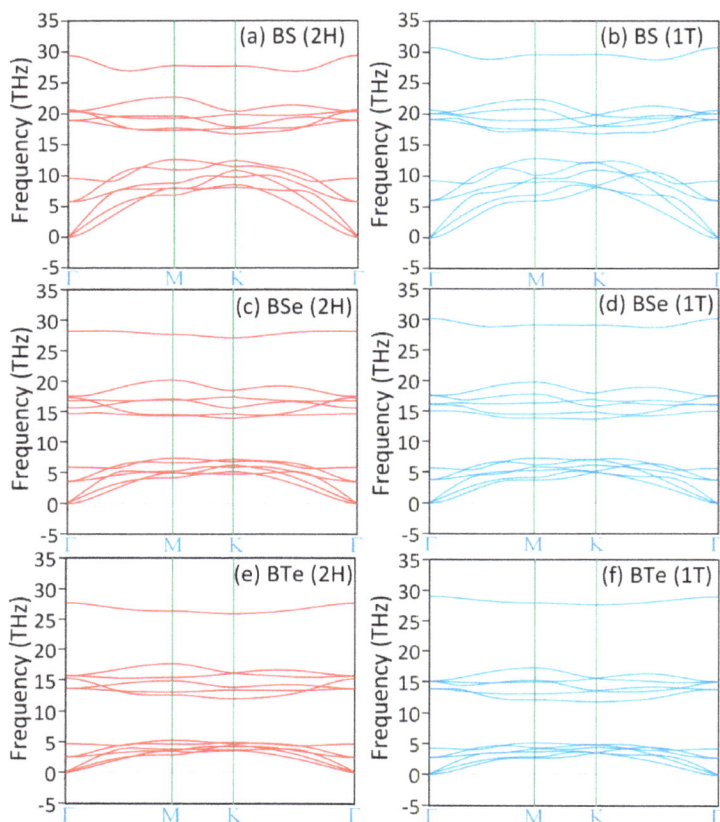

Figure 2. Phonon dispersions of free-standing and single-layer BX nanosheets.

Thermal stability at high temperatures is another critical property that is highly desirable for every material in engineering applications. We, therefore, next examined the thermal stability of single-layer boron monochalcogenides using the AIMD simulations at high temperatures. To this goal, AIMD calculations were conducted at different temperatures of 500 K, 1000 K, 1500 K, and 2000 K, for 15 ps long simulations. Snapshots of the BX monolayers with 2H and 1T phases after the 15 ps of AIMD simulations at high temperatures of 1500 K and 2000 K are illustrated in Figure 3. As a remarkable finding, all predicted BS and BSe monolayers were found to stay intact at the high temperature of 2000 K. As is clear, nevertheless, BTe monolayers with the both 2H and 1T lattices were disintegrated at 2000 K. The obtained results confirm the outstanding thermal stability of BX monolayers, which are all able to keep their structures intact at high temperatures, like 1500 K. As it was discussed earlier, 1T phases were found to be more energetically favourable since their lattice energies were slightly lower than those of 2H lattices. Higher energetic stability of 1T phases suggest that they are more probable to be experimentally realized, which is in agreement with the *r*-BS bulk layered structure with the well-known 1T atomic lattice [27]. Our results shown in Figure 3, confirm no sign of 2H to 1T phase transition at high temperatures, as 2H phases could keep their original hexagonal and graphene-like lattices intact. This finding not only reveals the remarkable thermal stability of BX nanomembranes with 1T and 2T phases, but also shows that the energy barriers for 2H to 1T phase transition in BX monolayers are considerably high, which cannot be passed by high temperature annealing.

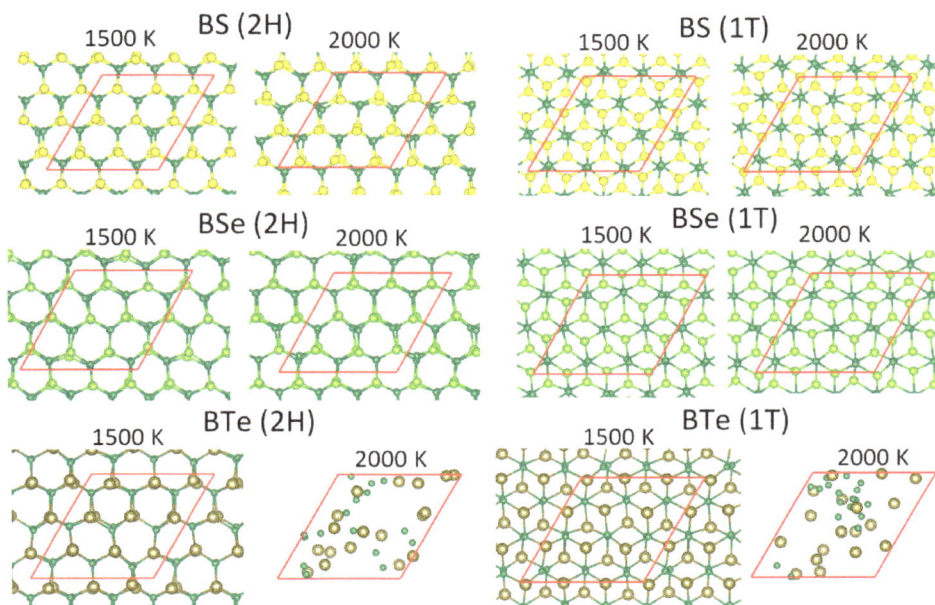

Figure 3. Snapshots of single-layer boron monochalcogenides at different temperatures of 1500 K and 2000 K after the AIMD simulations for 15 ps.

In Figure 4 band structures of BX monolayers along high symmetry directions predicted by PBE method are illustrated. Our results indicate that all boron monochalcogenides monolayers are wide and indirect band gap semiconductors. This observation suggests the limited application prospect of predicted monolayers as nanotransistors in post-silicon electronic. For the all BS monolayers valence band maximum (VBM) occurs around the Γ point. For the single-layer boron monochalcogenides with 1T atomic lattice, conduction band minimum (CBM) happens around the M point. In the case of BX monolayers with 2H atomic configuration, the CBM, however, occurs at M and K points. The band gaps of BS, BSe, and BTe monolayers with a 2H lattice within the PBE functional were measured to be 2.8 eV, 2.07 eV, and 1.49 eV, respectively. The corresponding values for 1T-BS, -BSe, and -BTe monolayers were found to be higher, 2.87 eV, 2.5 eV, and 1.7 eV, respectively.

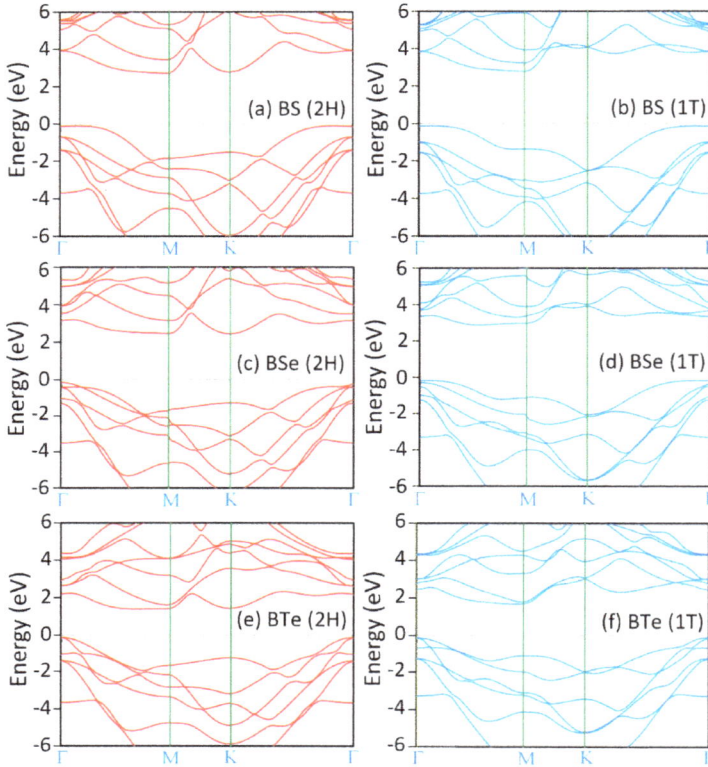

Figure 4. Band structure of single-layer BS, BSe, and BTe with 2H and 1T atomic lattices predicted by the PBE functional. The Fermi energy is aligned to zero.

Since the PBE method underestimates the band gap, we employed the HSE06 functional to more accurately report the band gap values in these 2D systems. For 2H-BS, -BSe, and -BTe, the band gaps were measured to be 3.85 eV, 2.96 eV, and 2.07 eV, respectively. In accordance with PBE results, the band gap of 1T lattices were found to be slightly higher, 4.0 eV, 3.43 eV, and 2.45 eV for single-layer BS, BSe, and BTe, respectively. As is clear, BS and BTe monolayers exhibit the largest and narrowest band-gaps, respectively.

Mechanical properties are among the most critical properties of any material for the design of nanodevices [44–54]. We finally study the mechanical properties of boron monochalcogenides predicted in this work. In Figure 5, the DFT predictions for the uniaxial stress-strain responses of single-layer BS, BSe, and BTe with 2H and 1T atomic lattices elongated along the armchair and zigzag directions are compared. For all of the predicted monolayers, the stress-strain response starts with an initial linear relation which is followed by a nonlinear trend up to the ultimate tensile strength point. The slope of the initial linear section of stress-strain curve is equal to the elastic modulus. In this work in order to report the elastic modulus, we fitted linear lines to the stress-strain values for the strain levels below 0.01. Within the elastic region, the strain along the traverse direction (ε_t) with respect to the loading strain (ε_l) is constant and can be used to evaluate the Poisson's ratio, using: $-\varepsilon_t/\varepsilon_l$. In Table 2, the mechanical properties of these 2D structures are summarized. Our results for the uniaxial loading along the armchair and zigzag directions show that linear sections coincide closely. As is also shown in Table 2, the elastic modulus along both considered loading directions are close, which suggest convincingly isotropic elasticity in the predicted BX monolayers. In contrast, the nonlinear sections of stress-strain responses are different depending on the loading direction. Interestingly, along the zigzag direction these systems show considerably higher tensile strength and stretchability as well. As is clear, the mechanical properties decline from BS to BTe. For the all BS monolayers, 2H lattices are stronger than 1T phases. Interestingly, the strain at ultimate tensile strength which is the representative of the stretchability, is very high for BS and BTe monolayers with 2H lattice for the loading along the zigzag direction. Notably, along the zigzag direction the strain at ultimate tensile strength of 2H-Bs and -BTe monolayers were found to be 0.26 and 0.3, respectively, comparable to that of the pristine graphene (~0.27 [55]) and hexagonal boron-nitride (~0.3 [56]).

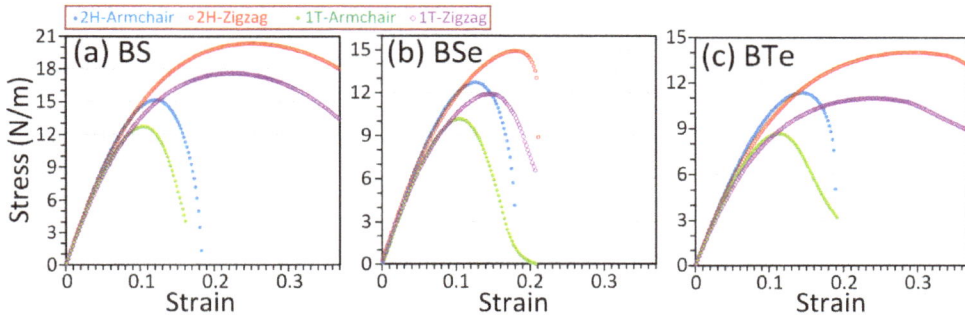

Figure 5. Uniaxial stress-strain responses of single-layer and free-standing boron monochalcogenides with 2H and 1T phases stretched along the armchair and zigzag directions.

Table 2. Mechanical properties of single-layer BS, BSe and BTe with 2H and 1T atomic lattices along the armchair and zigzag directions. Y, P, *UTS* and *SUTS* stand for elastic modulus, Poisson's ratio, ultimate tensile strength, and strain at ultimate tensile strength point, respectively. Stress units are in N/m.

Structure		$Y_{armchair}$	Y_{zigzag}	$P_{armchair}$	P_{zigzag}	$UTS_{armchair}$	UTS_{zigzag}	$SUTS_{armchair}$	$SUTS_{zigzag}$
BS	2H	203	201	0.13	0.12	15.17	20.31	0.12	0.26
	1T	195	193	0.13	0.12	12.69	17.60	0.11	0.23
BSe	2H	162	159	0.17	0.14	12.72	14.92	0.13	0.18
	1T	155	155	0.17	0.14	10.15	11.89	0.11	0.15
BTe	2H	131	130	0.17	0.14	11.36	14.01	0.15	0.3
	1T	122	121	0.17	0.14	11.00	8.66	0.12	0.27

4. Conclusions

In this work we introduced boron monochalcogenides as a class of 2D materials. Two different atomic lattices for BS, BSe, and BTe monolayers with stacking sequences of ABBA and ABBC were considered. According to first-principles density functional theory simulation results, all predicted monolayers were found to be dynamically stable. Ab initio molecular dynamics simulations reveal the remarkable thermal stability of all predicted monolayers at high temperatures over 1500 K. Electronic band structures confirm that all predicted single-layer boron monochalcogenides are indirect band gap semiconductors. According to the HSE06 method results, the band gaps of BS, BSe, and BTe monolayers with 2H atomic lattices were found to be 3.85 eV, 2.96 eV, and 2.07 eV, respectively. In this regard, the electronic band-gap of 1T lattices were found to be slightly higher, 4.0 eV, 3.43 eV, and 2.45 eV for single-layer BS, BSe, and BTe, respectively. Our first-principles results confirm convincingly isotropic elasticity and highly anisotropic tensile strengths in 2H and 1T boron monochalcogenide nanomembranes. It was found that along the zigzag direction these nanosheets show considerably higher tensile strength and stretchability as well. As an interesting finding, the stretchability of BS and BTe nanomembranes along the zigzag were found to be comparable to that of the graphene and hexagonal boron-nitride. This study introduces a class of stable and strong 2D materials with inherent semiconducting electronic characters and, thus, we are hopeful that the obtained result can motivate future experimental and theoretical studies to explore the possible synthesis routes and new physics and chemistry of these nanomembranes.

Author Contributions: B.M. initiated the project, conducted the all simulations and wrote the manuscript under the supervision of T.R.

Acknowledgments: Authors greatly acknowledge the financial support by European Research Council for COMBAT project (grant number 615132).

Conflicts of Interest: The authors declare no conflict of interest.

References

1. Novoselov, K.S.; Geim, A.K.; Morozov, S.V.; Jiang, D.A.; Zhang, Y.; Dubonos, S.V.; Grigorieva, I.V.; Firsov, A.A. Electric field effect in atomically thin carbon films. *Science* **2004**, *306*, 666–669. [CrossRef] [PubMed]
2. Geim, A.K.; Novoselov, K.S. The rise of graphene. *Nat. Mater.* **2007**, *6*, 183–191. [CrossRef] [PubMed]
3. Novoselov, K.S.; Mishchenko, A.; Carvalho, A.; Neto, A.H.C. 2D materials and van der Waals heterostructures. *Science* **2016**, *353*, aac9439. [CrossRef] [PubMed]
4. Mounet, N.; Gibertini, M.; Schwaller, P.; Campi, D.; Merkys, A.; Marrazzo, A.; Sohier, T.; Castelli, I.E.; Cepellotti, A.; Pizzi, G.; et al. Two-dimensional materials from high-throughput computational exfoliation of experimentally known compounds. *Nat. Nanotechnol.* **2018**, *13*, 246–252. [CrossRef] [PubMed]
5. Oganov, A.R.; Glass, C.W. Crystal structure prediction using ab initio evolutionary techniques: Principles and applications. *J. Chem. Phys.* **2006**, *124*, 244704. [CrossRef] [PubMed]
6. Glass, C.W.; Oganov, A.R.; Hansen, N. USPEX-Evolutionary crystal structure prediction. *Comput. Phys. Commun.* **2006**, *175*, 713–720. [CrossRef]
7. Oganov, A.R.; Lyakhov, A.O.; Valle, M. How evolutionary crystal structure prediction works-and why. *Acc. Chem. Res.* **2011**, *44*, 227–237. [CrossRef] [PubMed]
8. Mannix, A.J.; Zhou, X.F.; Kiraly, B.; Wood, J.D.; Alducin, D.; Myers, B.D.; Liu, X.; Fisher, B.L.; Santiago, U.; Guest, J.R.; et al. Synthesis of borophenes: Anisotropic, two-dimensional boron polymorphs. *Science* **2015**, *350*, 1513–1516. [CrossRef] [PubMed]
9. Feng, B.; Zhang, J.; Zhong, Q.; Li, W.; Li, S.; Li, H.; Cheng, P.; Meng, S.; Chen, L.; Wu, K. Experimental Realization of Two-Dimensional Boron Sheets. *Nat. Chem.* **2016**, *8*, 563–568. [CrossRef] [PubMed]
10. Zhou, X.F.; Dong, X.; Oganov, A.R.; Zhu, Q.; Tian, Y.; Wang, H.T. Semimetallic two-dimensional boron allotrope with massless Dirac fermions. *Phys. Rev. Lett.* **2014**, *112*, 085502. [CrossRef]
11. Zhang, Z.; Yang, Y.; Gao, G.; Yakobson, B.I. Two-Dimensional Boron Monolayers Mediated by Metal Substrates. *Angew. Chem. Int. Ed.* **2015**, *127*, 13214–13218. [CrossRef]
12. Kubota, Y.; Watanabe, K.; Tsuda, O.; Taniguchi, T. Deep ultraviolet light-emitting hexagonal boron nitride synthesized at atmospheric pressure. *Science* **2007**, *317*, 932–934. [CrossRef] [PubMed]
13. Song, L.; Ci, L.; Lu, H.; Sorokin, P.B.; Jin, C.; Ni, J.; Kvashnin, A.G.; Kvashnin, D.G.; Lou, J.; Yakobson, B.I.; et al. Large scale growth and characterization of atomic hexagonal boron nitride layers. *Nano Lett.* **2010**, *10*, 3209–3215. [CrossRef] [PubMed]
14. Mahmood, J.; Lee, E.K.; Jung, M.; Shin, D.; Choi, H.J.; Seo, J.M.; Jung, S.M.; Kim, D.; Li, F.; Lah, M.S.; et al. Two-dimensional polyaniline (C3N) from carbonized organic single crystals in solid state. *Proc. Natl. Acad. Sci. USA* **2016**, *113*, 7414–7419. [CrossRef] [PubMed]
15. Mahmood, J.; Lee, E.K.; Jung, M.; Shin, D.; Jeon, I.Y.; Jung, S.M.; Choi, H.J.; Seo, J.M.; Bae, S.Y.; Sohn, S.D.; et al. Nitrogenated holey two-dimensional structures. *Nat. Commun.* **2015**, *6*, 6486. [CrossRef] [PubMed]
16. Thomas, A.; Fischer, A.; Goettmann, F.; Antonietti, M.; Müller, J.O.; Schlögl, R.; Carlsson, J.M. Graphitic carbon nitride materials: Variation of structure and morphology and their use as metal-free catalysts. *J. Mater. Chem.* **2008**, *18*, 4893–4908. [CrossRef]
17. Algara-Siller, G.; Severin, N.; Chong, S.Y.; Björkman, T.; Palgrave, R.G.; Laybourn, A.; Antonietti, M.; Khimyak, Y.Z.; Krasheninnikov, A.V.; Rabe, J.P.; et al. Triazine-based graphitic carbon nitride: A two-dimensional semiconductor. *Angew. Chem.* **2014**, *53*, 7450–7455. [CrossRef] [PubMed]

18. Aufray, B.; Kara, A.; Vizzini, S.; Oughaddou, H.; Leandri, C.; Ealet, B.; Le Lay, G. Graphene-like silicon nanoribbons on Ag(110): A possible formation of silicene. *Appl. Phys. Lett.* **2010**, *96*, 183102. [CrossRef]
19. Vogt, P.; De Padova, P.; Quaresima, C.; Avila, J.; Frantzeskakis, E.; Asensio, M.C.; Resta, A.; Ealet, B.; Le Lay, G. Silicene: Compelling experimental evidence for graphenelike two-dimensional silicon. *Phys. Rev. Lett.* **2012**, *108*, 155501. [CrossRef] [PubMed]
20. Bianco, E.; Butler, S.; Jiang, S.; Restrepo, O.D.; Windl, W.; Goldberger, E.J. Stability and exfoliation of germanane: A germanium graphane analogue. *ACS Nano* **2013**, *7*, 4414–4421. [CrossRef] [PubMed]
21. Geim, A.K.; Grigorieva, I.V. Van der Waals heterostructures. *Nature* **2013**, *499*, 419–425. [CrossRef] [PubMed]
22. Wang, Q.H.; Kalantar-Zadeh, K.; Kis, A.; Coleman, J.N.; Strano, M.S. Electronics and optoelectronics of two-dimensional transition metal dichalcogenides. *Nat. Nanotechnol.* **2012**, *7*, 699–712. [CrossRef] [PubMed]
23. Radisavljevic, B.; Radenovic, A.; Brivio, J.; Giacometti, V.; Kis, A. Single-layer MoS_2 transistors. *Nat. Nanotechnol.* **2011**, *6*, 147–150. [CrossRef] [PubMed]
24. Das, S.; Demarteau, M.; Roelofs, A. Ambipolar phosphorene field effect transistor. *ACS Nano* **2014**, *8*, 11730–11738. [CrossRef] [PubMed]
25. Li, L.; Yu, Y.; Ye, G.J.; Ge, Q.; Ou, X.; Wu, H.; Feng, D.; Chen, X.H.; Zhang, Y. Black phosphorus field-effect transistors. *Nat. Nanotechnol.* **2014**, *9*, 372–377. [CrossRef] [PubMed]
26. Bandurin, D.A.; Tyurnina, A.V.; Geliang, L.Y.; Mishchenko, A.; Zólyomi, V.; Morozov, S.V.; Kumar, R.K.; Gorbachev, R.V.; Kudrynskyi, Z.R.; Pezzini, S.; et al. High electron mobility, quantum Hall effect and anomalous optical response in atomically thin InSe. *Nat. Nano* **2017**, *12*, 223–227. [CrossRef] [PubMed]
27. Cherednichenko, K.A.; Kruglov, I.A.; Oganov, A.R.; le Godec, Y.; Mezouar, M.; Solozhenko, V.L. Boron monosulfide: Equation of state and pressure-induced phase transition. *J. Appl. Phys.* **2018**, *123*, 135903. [CrossRef]
28. Zólyomi, V.; Drummond, N.D.; Fal'Ko, V.I. Electrons and phonons in single layers of hexagonal indium chalcogenides from ab initio calculations. *Phys. Rev. B Condens. Matter Mater. Phys.* **2014**, *89*, 205416. [CrossRef]
29. Fan, D.; Yang, C.; Lu, S.; Hu, X. Two-Dimensional Boron Monosulfides: Semiconducting and Metallic Polymorphs. *arXiv* **2018**, arXiv:1803.03459.
30. Kresse, G. From ultrasoft pseudopotentials to the projector augmented-wave method. *Phys. Rev. B* **1999**, *59*, 1758–1775. [CrossRef]
31. Kresse, G.; Furthmüller, J. Efficiency of ab-initio total energy calculations for metals and semiconductors using a plane-wave basis set. *Comput. Mater. Sci.* **1996**, *6*, 15–50. [CrossRef]
32. Kresse, G.; Furthmüller, J. Efficient iterative schemes for ab initio total-energy calculations using a plane-wave basis set. *Phys. Rev. B* **1996**, *54*, 11169–11186. [CrossRef]
33. Perdew, J.; Burke, K.; Ernzerhof, M. Generalized Gradient Approximation Made Simple. *Phys. Rev. Lett.* **1996**, *77*, 3865–3868. [CrossRef] [PubMed]
34. Blöchl, P.E.; Jepsen, O.; Andersen, O.K. Improved tetrahedron method for Brillouin-zone integrations. *Phys. Rev. B* **1994**, *49*, 16223–16233. [CrossRef]
35. Monkhorst, H.; Pack, J. Special points for Brillouin zone integrations. *Phys. Rev. B* **1976**, *13*, 5188–5192. [CrossRef]
36. Krukau, A.V.; Vydrov, O.A.; Izmaylov, A.F.; Scuseria, G.E. Influence of the exchange screening parameter on the performance of screened hybrid functionals. *J. Chem. Phys.* **2006**, *125*, 224106. [CrossRef] [PubMed]
37. Togo, A.; Oba, F.; Tanaka, I. First-principles calculations of the ferroelastic transition between rutile-type and $CaCl_2$-type SiO_2 at high pressures. *Phys. Rev. B Condens. Matter Mater. Phys.* **2008**, *78*, 134106. [CrossRef]
38. Lin, Y.-C.; Dumcenco, D.O.; Huang, Y.-S.; Suenaga, K. Atomic mechanism of the semiconducting-to-metallic phase transition in single-layered MoS_2. *Nat. Nanotechnol.* **2014**, *9*, 391–396. [CrossRef] [PubMed]
39. Mattheiss, L.F. Band structures of transition-metal-dichalcogenide layer compounds. *Phys. Rev. B* **1973**, *8*, 3719–3740. [CrossRef]
40. Wypych, F.; Schöllhorn, R.; Schollhorn, R.; Schöllhorn, R. 1T-MoS_2, a new metallic modification of molybdenum disulfide. *J. Chem. Soc. Chem. Commun.* **1992**, *19*, 1386–1388. [CrossRef]
41. Momma, K.; Izumi, F. VESTA 3 for three-dimensional visualization of crystal, volumetric and morphology data. *J. Appl. Crystallogr.* **2011**, *44*, 1272–1276. [CrossRef]
42. Silvi, B.; Savin, A. Classification of Chemical-Bonds Based on Topological Analysis of Electron Localization Functions. *Nature* **1994**, *371*, 683–686. [CrossRef]

43. Henkelman, G.; Arnaldsson, A.; Jónsson, H. A fast and robust algorithm for Bader decomposition of charge density. *Comput. Mater. Sci.* **2006**, *36*, 354–360. [CrossRef]

44. Sadeghzadeh, S. Borophene sheets with in-plane chain-like boundaries; a reactive molecular dynamics study. *Comput. Mater. Sci.* **2018**, *143*, 1–14. [CrossRef]

45. Sadeghzadeh, S. The creation of racks and nanopores creation in various allotropes of boron due to the mechanical loads. *Superlattices Microstruct.* **2017**, *111*, 1145–1161. [CrossRef]

46. Le, M.-Q.; Batra, R.C. Mode-I stress intensity factor in single layer graphene sheets. *Comput. Mater. Sci.* **2016**, *118*, 251–258. [CrossRef]

47. Ganji, A.R.P.; Armat, M.R.; Tabatabaeichehr, M.; Mortazavi, H. The Effect of Self-Management Educational Program on Pain Intensity in Elderly Patients with Knee Osteoarthritis: A Randomized Clinical Trial. *Open Access Maced. J. Med. Sci.* **2018**. [CrossRef]

48. Mortazavi, H. Designing a multidimensional pain assessment tool for critically Ill elderly patients: An agenda for future research. *Indian J. Crit. Care Med.* **2018**, *22*, 390–391. [CrossRef]

49. Le, M.-Q. Reactive molecular dynamics simulations of the mechanical properties of various phosphorene allotropes. *Nanotechnology* **2018**, *29*, 195701. [CrossRef] [PubMed]

50. Nguyen, D.T.; Le, M.Q.; Nguyen, V.T.; Bui, T.L. Effects of various defects on the mechanical properties of black phosphorene. *Superlattices Microstruct.* **2017**, *112*, 186–199. [CrossRef]

51. Shirazi, A.H.N.; Abadi, R.; Izadifar, M.; Alajlan, N.; Rabczuk, T. Mechanical responses of pristine and defective C3N nanosheets studied by molecular dynamics simulations. *Comput. Mater. Sci.* **2018**, *147*, 316–321. [CrossRef]

52. Abadi, R.; Shirazi, A.H.N.; Izadifar, M.; Sepahi, M.; Rabczuk, T. Fabrication of nanopores in polycrystalline boron-nitride nanosheet by using Si, SiC and diamond clusters bombardment. *Comput. Mater. Sci.* **2018**, *145*, 280–290. [CrossRef]

53. Shi, L.B.; Zhang, Y.Y.; Xiu, X.M.; Dong, H.K. Structural Characteristics and Strain Behaviors of Two-Dimensional C3N: First Principles Calculations. *Carbon* **2018**, *134*, 103–111. [CrossRef]

54. Mortazavi, H.; Tabatabaeichehr, M.; Taherpour, M.; Masoumi, M. Relationship Between Home Safety and Prevalence of Falls and Fear of Falling Among Elderly People: A Cross-Sectional Study. *Mater. Soc. Med.* **2018**, *30*, 103–107. [CrossRef]

55. Liu, F.; Ming, P.; Li, J. Ab initio calculation of ideal strength and phonon instability of graphene under tension. *Phys. Rev. B Condens. Matter Mater. Phys.* **2007**, *76*, 064120. [CrossRef]

56. Peng, Q.; Ji, W.; De, S. Mechanical properties of the hexagonal boron nitride monolayer: Ab initio study. *Comput. Mater. Sci.* **2012**, *56*, 11–17. [CrossRef]

energies

MDPI

Article

Finite Element Analysis to the Effect of Thermo-Mechanical Loads on Stress Distribution in Buried Polyethylene Gas Pipes Jointed by Electrofusion Sockets, Repaired by PE Patches

Reza Khademi-Zahedi [1,*] and Pouyan Alimouri [2]

[1] Institute of Structural Mechanics, Bauhaus-Universität Weimar, 99423 Weimar, Germany
[2] Department of Mechanical Engineering, Shahid Chamran University of Ahvaz, Ahvaz, Iran;
Alimouri_P@yahoo.com
* Correspondence: reza.khademi.zahedi@uni-weimar.de

Received: 7 September 2018; Accepted: 8 October 2018; Published: 19 October 2018

Abstract: Polyethylene (PE) gas pipes can be jointed together by electrofusion PE fittings, which have sockets that are fused onto the pipe. Additionally, electrofused PE patches can be used to repair defected pipes. When these pipelines are buried under the ground, they can experience sever local stresses due to the presence of pipe joints, which is superimposed on the other effects including the soil-structure interaction, traffic load, soil's column weight, a uniform internal pressure, and thermal loads imposed by daily and/or seasonal temperature changes. The present contribution includes two cases. At first, stress variations in buried polyethylene gas pipe and its socket due to the aforementioned loading condition is estimated using finite element. The pipe is assumed to be made of PE80 material and its jointing socket material is PE100. Afterward, the effects of aforementioned thermo-mechanical loads on the stress distribution in patch repaired buried pipes are well investigated. The soil physical properties and the underground polyethylene pipe installation method are based on the American association of state highway and transportation officials and American society for testing and material standards. The computer simulation and analysis of stresses are performed through the finite element package of ANSYS Software. Stress concentrations can be observed in both components due to the presence of the socket or the repair patch. According to the results, the electrofusion sockets can be used for joining PE gas pipes even in hot climate areas. The maximum values of these stresses happen to be in the pipe. Also, the PE100 socket is more sensitive to a temperature drop. Additionally, all four studied patch arrangements show significant reinforcing effects on the defected section of the buried PE gas pipe to withstand applied loads. Meanwhile, the defected buried medium density polyethylene (MDPE) gas pipe and its saddle fused patch can resist the imposed mechanical and thermal loads of +22 °C temperature increase.

Keywords: buried gas distribution pipes; electrofusion socket joints; patch repair; medium density polyethylene (MDPE); high density polyethylene (HDPE); Von Mises stress; finite element method; temperature variation

---

1. Introduction

Petroleum, natural gas, and condensates are naturally occurring substances which are discovered within the Earth's crust, are thought to originate from decomposed animal and plant matter. Scientists believe the plants and animals died in the distant past, and were gradually buried by thick layers of sediments. Over a long period of time, and with pressure and temperature, the organic materials were transformed into the oil and gas which are found today. When oil and gas are removed from the ground they are sent to refineries by pipelines. Then, many products from these materials, which potentially contain several chemicals called hydrocarbons, can be obtained including energy for power, motor oil, gasoline for cars, diesel fuel for trucks and trains, hi-octane fuels for planes, heating oil for houses etc. Several other materials also come from petroleum such as plastics, asphalt, grease, lubricating oil, materials for clothes, chemicals for everyday use, paints etc. Since oil has natural gas in it, when oil is produced often some gas is produced with the oil if natural gas is in the solution. The majority of oil is trapped in the tiny pore spaces between grains of rock or sand. Oil and gas are discovered in natural traps which consist of domes or faults within the earth. Impermeable rock above the trap that fluid cannot pass through it stops the oil and gas from move up to the surface. Without traps, the oil and gas could migrate all the way to the surface and evaporate. Figure 1 shows an example of an oil reservoir with a gas cap.

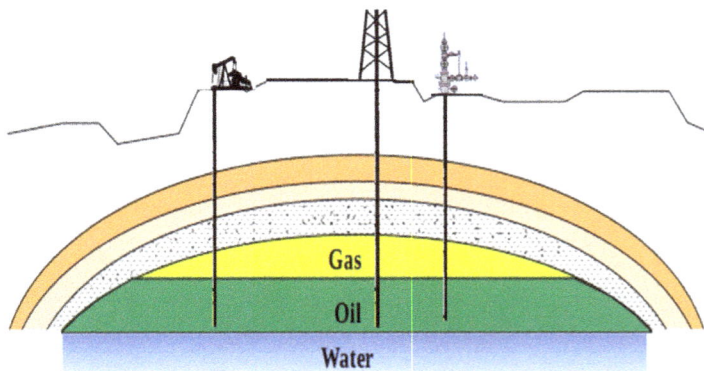

Figure 1. An example of an oil reservoir with a gas cap.

After geophysicists find reservoirs and process the data to construct pictures of what the earth looks like underground, drilling companies start to drill into the proposed reservoirs. A drilling rig is a package of special equipment put together that enables us to create a hole to a projected depth for producing oil or natural gas [1–4]. Finally, the well is produced into a pipeline, which takes it to production facilities on surface. The production facilities on surface separate out the gas, oil, and water into their separate phases. From there, the oil and gas may be refined further before being ready to market. Finally, the gas and oil can be sold to power cars and heat houses.

High density (HD) and medium density (MD) polyethylene materials are often used to produce gas pipes. A pipeline network needs a long length pipe to dispatch natural gas. It is troublesome to produce long pipes, hence the needed lengths of pipe is assembled by connecting short pipes together. Additionally, in order to construct long pipe networks, it is essential to utilize joining methods such as butt-welding, electro-fusion, mechanical joining etc. Mainly, electro-fusion joint is extensively employed since it is empirically feasible to perform even in areas where access is restricted. Generally, buried pipes are divided into two categories: flexible pipes, where the pipe deforms under applied loads at least 2 percent of the pipe diameter without any effect to the pipe material, and rigid pipes, where the pipe structure changes (for example with crack creation) in the mentioned deformation [5]. PE pipes behave as flexible pipes when installed underground. In the underground application of PE pipes, the installation procedure and the design method are of great importance for natural gas distributor companies. The first step in buried pipe design is to determine the soil dead load and surcharge loads on the pipe and ensure the safe operation of the PE pipes for a long period of time. The load resulted from soil column weight, vehicle wheel, pipe internal pressure, and daily or seasonal temperature changes which induced horizontal and vertical deflections along with significant stresses in the pipe wall. Consequently the soil showed reaction to the aforementioned pipe action and will limit the pipe deflection. Several experimental and analytical methods have been proposed by researchers to calculate the applied loads, deflections, and stresses on the buried pipes that, although suitable for most underground applications, do not produce accurate results [6]. The modern buried-pipe design was proposed by Marston from Iowa University in the early 20thcentury [6,7]. Spangler continued theoretical studies on flexible buried pipes by assuming soil and pipe as linear elastic materials and proposed Iowa's modified formula to predict deflection in flexible buried pipes [7]. Afterward flexible pipe materials including aluminum, polyvinyl, and polyethylene were used in underground applications to improve pipe-soil behavior and reduce costs. Pipe-soil interaction is a combination system that significant share of the applied load is carried by the soil around the pipe which makes it difficult to calculate the induced stresses in the pipe material. One of the strongest numerical approaches to investigate underground structure responses to the applied loads is the finite element method. Abaqus and Ansys software has been used by several researchers for three-dimensional modeling of the aforementioned structures. In this research, the methods discussed in contributions on finite element modeling of buried polyethylene gas pipes which can be found in references [8–13] are used to further investigate the stress distribution in special cases of buried polyethylene gas pipe problems which are subjected to the local geometry changes compared to simple buried pipes. This study can be divided in two parts. In the first part, the case of joining buried PE pipes by electrofusion socket joints is well discussed. In the second part of the research, four different patch arrangements used to repair defected buried pipes are investigated, and their reinforcing effects on damaged parts are discussed. The pipe, socket/patch, the soil grades surrounding them, and the underground installation procedure and burial depth are all selected based on the standards commonly used in gas distribution companies. For both cases, the simultaneous effects of mechanical loads including pipe internal pressure, vehicle traffic load, soil column load, and more critical thermal loads of temperature variations in PE pipe and its socket/patch on the induced stresses in PE80 pipe and PE100 socket/patch materials are investigated using Ansys software. In both cases when the model is imposed to the aforementioned thermo-mechanical loads, stress concentrations will appear in the pipe and its socket/patch due to local changes in the pipe geometry.

2. Basic Design Theory

In engineering design, soil classification systems for structural applications are based on soil mechanical and physical properties including grain sizes. Commonly, AASHTO and ASTM standards are used as guides for these purposes [5,14]. The investigation of structural behavior of underground systems starts with the determination of applied loads. The design of underground structures is in accordance with the principle of soil-pipe interaction. Estimating the loads on structures, including gas pipelines used in underground space, depends on the pipe installation procedure in the trench that can be found in ASTM D 2321 standard [15,16]. The mechanical properties of the backfill environment or embedment should meet the special structural application. The mechanical properties of the soil grades used for bedding, haunching, initial and final backfill and also asphalt cover, along with installation dimensions are selected based on references [8–11]. Additionally similar loading conditions are introduced and imposed.

3. Stress Investigation in Socket Joint of Buried Polyethylene Gas Pipe

The socket material for joining MDPE buried gas pipes was selected from the PE100. Based on references [8–11] PE80 material was selected for the pipe. In order to investigate the stress distribution in a pressurized PE gas pipe, the value of 11 was selected for pipe and its socket SDR (Standard dimension ratio) Geometrical and mechanical properties of PE100 socket material are presented in Table 1 [17].

Table 1. Geometrical and mechanical properties of PE100 socket.

	Geometrical Properties		Mechanical Properties				
SDR	Outer Diameter (mm)	Wall Thickness (mm)	Modulus of Elasticity	Yield Strength (MPa)	Rupture Strength (MPa)	Poisson's Ratio	Thermal Expansion Coefficient ($\frac{1}{°C}$)
11	110	10	1034	24.8	35.1	0.4	0.0002

The type of the joining is electro fused socket where the related dimensions are presented in Figure 2.

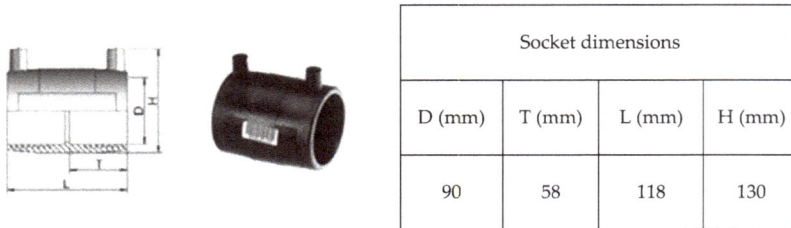

Socket dimensions			
D (mm)	T (mm)	L (mm)	H (mm)
90	58	118	130

Figure 2. Selected socket geometry and dimensions for the investigated PE pipe [18]. D: Socket inside diameter; T: Inner length; L: Socket length; H: Socket height.

Generally, PE100 material is the proper option to join PE80 pipes. PE100 material has the minimum allowable stress of 10 MPa for a working life of 50 years in 20 °C design temperature. In contrast, the minimum allowable stress for PE80 material for a working life of 50 years in 20 °C design temperature is as low as 8 MPa. PE100 material shows better resistance against perforation than PE80 material. Additionally, PE100 material shows better resistance to fast crack propagation in comparison to PE80 material. This means if a crack initiates in a PE100 material, it will stop propagating in a short length. This advantage is of great importance especially for pipes loaded under high pressure values that fast crack propagation is of magnificent importance. PE100 material shows good behavior against creep rupture compared to PE80 material. Additionally, for two different PE100 and PE80 pipes with the same SDR, PE100 pipe material can be used for higher pressure values. That means with the same inside pressure for PE100 and PE80 pipe material, the wall thickness of PE100 pipe material can be chosen 19% less than that of PE80 pipe material [19].

To investigate the stress distribution in the socket the simultaneous effects of soil column weight, traffic wheel load, pipe and socket temperature changes, internal pressure are considered and implemented in ANSYS (V12, ANSYS, Pittsburgh, PA, USA) software. Figure 3 illustrates a typical finite element model of pipe and socket subjected to internal pressure. It should be noted that the dimension of the model upper surface is $x = 1.5$ m and $z = 1.5$ m. Furthermore, because of the identical effect of the concentrated and distributed wheel load in the selected installation depth, the effect of distributed wheel load on the pipe and its socket is investigated. Additionally the soil-pipe interaction is also incorporated in the model by using face-to-face contact elements (conta 172 and target 170).

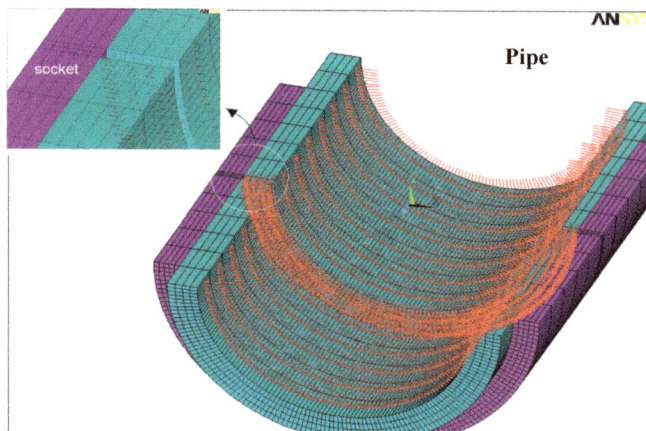

Figure 3. Typical finite element model of pipe and socket subjected to internal pressure.

4. Evaluation of the Results

As previously mentioned, the effect of traffic load, soil weight, temperature changes, internal pressure, and the presence of the local changes in the pipe geometry were all included to investigate their overall effects on the stress distribution in the buried PE80 pipe and its related socket joint from PE100 material. For the following section of this research, the value of 1.5 is selected for the design coefficient (factor of safety). Considering the operating temperature of 35 °C for underground pipes buried at a depth of 125 cm, based on climate conditions in hot areas and Ahvaz city [8–13,20], according to the reported minimum strength value for polyethylene PE100 material [21], the design stress (maximum allowable stress) for this research is 5.4 MPa. Additionally, this value of design stress is calculated to be 4.3 MPa for the PE80 pipe material. In all stages of the next section of this investigation, the mentioned values are considered as the maximum applicable stresses on the pipe.

4.1. Results of Finite Element Solution for the Socket Jointed Buried Pipe

In this section, various thermal and mechanical loadings applied to the buried socket jointed polyethylene gas pipe during the operation period are considered to be software inputs, and the resulting induced stresses are evaluated and discussed using finite element method.

The geometry and the installation procedure of the socket jointed polyethylene pipe are similar to that of normally buried polyethylene pipe which is well described [8–12]. Therefore, the same temperature variations for the buried pipe impose to the socket too.

The finite element results for maximum values of Von Mises and axial stresses under the simultaneous effects of the soil column weight, pipe internal pressure, vehicle wheel load and temperature variations of +5 °C, −15 °C and 0 °C (no variation) in the buried pipe and its socket are calculated and presented in Table 2. The maximum axial stress and the maximum Von Mises stress values in the wall of buried socket joint of the gas distribution pipe under the simultaneous effect of mechanical loads including soil load, internal pressure, vehicle wheel load and thermal loads in the form of maximum temperature drop of 15 °C and increase of 5 °C are calculated and plotted in Figure 4. For Figure 4, the coordinate system is selected so that the origin of the horizontal axes represents the socket internal surface and the stress variations are plotted from the origin to the socket crown. As expected, the upper curves pertain to the temperature decrease. Additionally, the dashed curve indicates the allowable stress value for design temperature of 35 °C based on a 50-year life expectancy for the pipe and socket material. Additionally the Von Mises stresses variations are plotted from the socket internal surface to its crown.

Table 2. The simultaneous effect of soil column pressure, pipe internal pressure, vehicle wheel load, and temperature variations.

Item	Maximum Stress (MPa)	$\Delta T = 0\,^\circ C$	$\Delta T = +5\,^\circ C$	$\Delta T = -15\,^\circ C$
Pipe	Axial stress	0.56	0.21	1.86
	Von Mises stress	2.19	2.30	2.45
Socket	Axial stress	2.52	1.74	4.14
	Von Mises stress	2.6	2.11	4.5

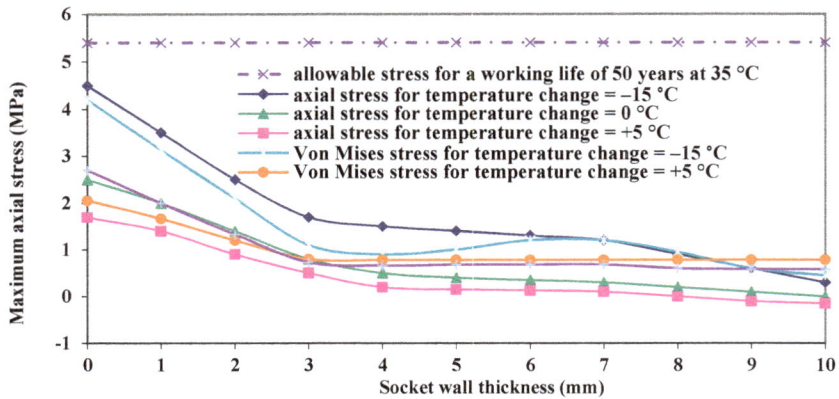

Figure 4. Maximum axial and Von Mises stress variations in the wall of buried socket joint (the simultaneous effect of soil load, internal pressure, vehicle wheel load and temperature variations).

Figure 5 shows the simultaneous effect of the aforementioned mechanical and thermal loads on the maximum axial and the maximum Von Mises stresses along the socket length of the buried polyethylene gas pipe. The origin of the horizontal axis is supposed to be the socket center. As the result of the geometry and loading symmetry, only the stress values of half of the socket are presented. Similarly, the dashed line is for the allowable stress values for the material of PE100. Additionally, maximum axial stress variations and maximum Von Mises stress variations along the half of pipe length under the simultaneous effect of soil load, internal pressure, and vehicle wheel load and temperature variations are calculated and plotted as Figure 6.

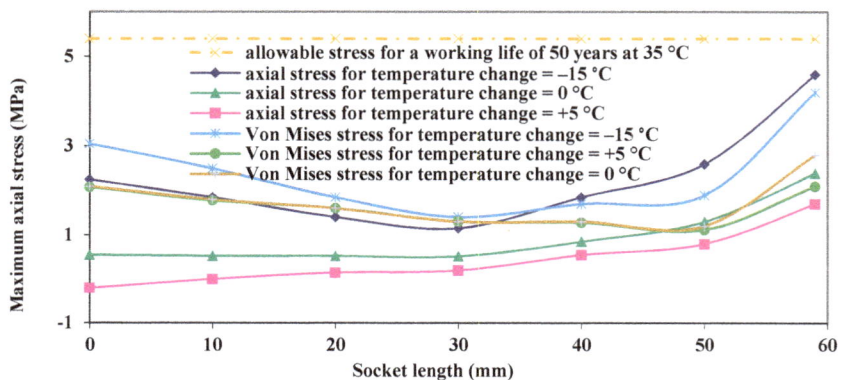

Figure 5. Maximum axial and Von Mises stress variations along the socket length (the simultaneous effect of soil load, internal pressure, vehicle wheel load and temperature variations).

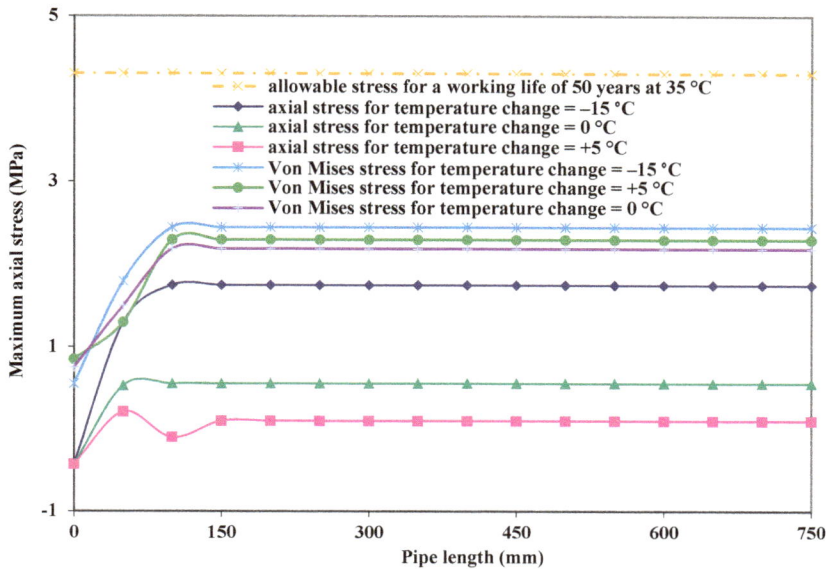

Figure 6. Maximum axial and Von Mises stress variations along the pipe length (the simultaneous effect of soil load, internal pressure, vehicle wheel load and temperature variations).

Considering the presented maximum stress results in Table 2 and Figures 4–6 it can be concluded that:

1. A daily temperature rise of +5 °C decreases the maximum axial stresses in the socket joint from 2.52 MPa to 1.74 MPa which is about 31% and additionally decreases the maximum axial stresses in the pipe from 0.56 MPa to 0.21 MPa which is about 63%.
2. A daily temperature rise +5 °C decreases maximum Von Mises stresses in the socket joint from 2.6 MPa to 2.11 MPa which is about 19% and additionally increases maximum Von Mises stresses in the pipe from 2.19 MPa to 2.30 MPa which is about 3.8%.
3. A daily temperature drop of −15 °C increases maximum axial stresses in the socket joint from 2.52 MPa to 4.14 MPa which is about 64% and additionally increases maximum axial stresses in the pipe from 0.56 MPa to 1.86 MPa which is about 235%.
4. A daily temperature drop of −15 °C increases maximum Von Mises stresses in the socket joint from 2.60 MPa to 4.50 MPa which is about 73% and additionally increases maximum Von Mises stresses in the pipe from 2.19 MPa to 2.45 MPa which is about 12%.
5. According to the Figures 4–6, in all cases the maximum values for both axial and Von Mises stresses occurs at the middle of the socket internal surface while the minimum values of the aforementioned stresses in the socket occur where the internal surface of the socket joins the pipe outer surface.

After investigating the effect of socket joints, in the next section the effect of patch repair is well studied by finite element method.

4.2. Results of Finite Element Solution for the Patch Repaired Pipe

Engineering structures bear or transfer several loads and may be defected during their lifetime due to several reasons. An understanding of fracture mechanisms plays an important role in structural materials safe design. Therefore several numerical methods have been used to investigate and model fracture in materials [22–40]. The calculation of stress distribution in a damaged underground polyethylene gas pipe using finite element modeling was well discussed in [9], where several circular and elliptical shaped defects with various circular hole diameter and ellipse hole diameter ratios a/b, were modeled at the pipe crown. Additionally, the effect of 4 types of polyethylene patch repair arrangements on the stress reduction in the aforementioned defected pipes was well investigated by the computational modeling and analysis tools of ANSYS software [10]. In the following, we continued the study in references [9,10] to investigate the buried PE pipe with more details and under more critical loading conditions. The finite element code is used for modeling and analysis of the pipe, patch and their surrounding and since stress distribution in the pipe wall is very important, 3D brick-type SOLID95 elements, available in the ANSYS software package elements library, are well employed to model the physical medium. The SOLID95 element has three degrees of freedom at each node (translations in the x, y, and z coordinate directions) that makes it very suitable for solving curved boundary problems. Additionally, to have the best mesh control with a minimal number of finite elements and to reduce the computational time and costs, the mapped (structured) finite element mesh which typically has a regular pattern, is performed to control the mesh efficiency. In this research we study the stress reduction effects on defected polyethylene gas pipes reinforced by polyethylene patches under sever thermo-mechanical loads for four various patch arrangements including semi-cylindrical, circular-partial, square-partial and saddle fusion patches to find the optimum patch shape. For the purpose of finding an appropriate patch shape and geometry, identical thickness (4.763 mm), inside diameter (114.3 mm) and material properties (PE 100) are considered for all patch types. It is assumed that the pipe and patch materials behave as linear elastic and have isotropic properties. A typical pipe-patch arrangement (saddle fusion patch) and related dimensions can be found on Figure 7. Additionally, the pipe and patch material mechanical properties can be found on Table 3.

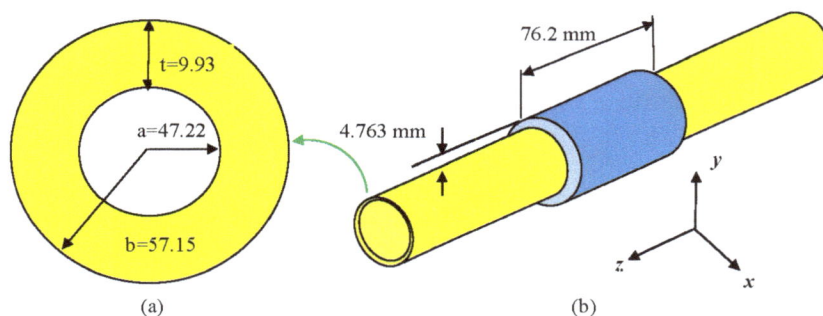

Figure 7. Polyethylene pipe and patch arrangement. (**a**) Two dimensional view (dimensions in mm) of the cross section of the medium density polyethylene gas pipe. (**b**) Three dimensional model of patch repaired pipe showing patch dimensions [8–12,41].

Table 3. Mechanical properties of the PE80 pipe and PE100 patch materials [8–12,41].

Material	Mechanical Properties			
	Elasticity Module (MPa)	Yield Stress (MPa)	Poisson Ratio	Thermal Expansion Coefficient ($\frac{1}{°C}$)
PE80	427	19.3	0.35	0.0002
PE100	1034	24.8	0.4	0.0002

In the next section we investigate the aforementioned models with more details by superimposing various thermal loads in the form of temperature variations to the mechanical loads including surcharge loads, soil column weight, soil-pipe interaction, and inside pressure of 4 bar applied to the repaired defected polyethylene pipe and its patch. The results of previous section show that the maximum Von Mises stress values are well below the allowable stresses (based on the design factor of 1.5) for working life of 50 years at 35 °C. Additionally lower values of safety factor (design factor) are applicable to design underground gas pipes. By selecting the value of 1.25 for safety factor, the values of allowable stresses for working life of 50 years at 35 °C will be 5.2 MPa and 6.5 MPa for PE80 pipe and PE100 patch materials respectively. We prefer to use these values for the following section of the research.

Before starting to investigate the complicated case of underground problem and soil-structure interaction for buried patch repaired polyethylene gas pipe, we perform verification by finite element modeling of the patch repaired PE pipe not buried under the ground and only loaded to a final internal pressure of 4 bars. The results are well described in the following section.

4.2.1. Finite Element Modeling and Solution for an Unburied Patch Repaired Pipe with Circular Wall Holes

In this section, we use finite element modeling to evaluate stress distribution in a defected unburied MDPE gas pipe that is repaired by applying a 76.2 mm (3") long saddle fusion patch to verify the applicability and accuracy of the computer simulation to solve patch repaired pipe problems and to find the optimum finite element model dimension for pipe length. Therefore, a 4-inch (nominal diameter) unburied MDPE pipe and its related patch, which is subjected only to a final pressure of 405,300 Pa (4 bar), is modeled computationally. The modeled pipe had circular holes with various diameters (5, 7.5, 10, 12.5, 15, 17.5, 20 mm) at the pipe crown, which are repaired by a saddle fusion patch made of high-density polyethylene material. It is assumed that the related patch is electro fused to the pipe, which creates a strong continuous connection. A geometric computer model of the pipe and its related patch was created in ANSYS so that the center of the coordinate system aligned with the center of the pipe and its patch and the z-axis aligned with the axis of the pipe and its patch. Also, the appropriate boundary conditions selected for the finite elements in this section is as Equations (1) and (2):

$$@ \ z = \pm \frac{l}{2}; \quad u_z = 0 \tag{1}$$

$$@ \ x = 0; \quad u_x = 0 \tag{2}$$

The plane strain constraint in the direction of the pipe axis (z-axis) is assumed for the long pipe where this constraint is applied to the nodes located at the related coordinates as seen in Equation (1). Additionally, Equation (2) is well applied because of the model and loading condition symmetry in y-z plane.

Unburied Patch Repair Finite Element Model Evaluation

The good performance of the present method is demonstrated in Table 4 by a number of numerical examples. Maximum values of hoop and Von Mises stresses in pipe and its related patch for different hole diameters and pipe lengths derived by the finite element method are presented.

Table 4. Comparing maximum stress values in pipe and patch under internal pressure, investigating the pipe length size.

Model Dimension			Maximum Stress in Pipe		Maximum Stress in Patch	
Hole Diameter (mm)	Pipe Length (m)	Number of Elements	Von Mises Stress (MPa)	Hoop Stress (MPa)	Von Mises Stress (MPa)	Hoop Stress (MPa)
	0.5	17,520	2.94	2.88	3.4	3.75
5		18,889	2.93	2.87	3.39	3.74
	1.5	24,732	2.92	2.86	3.4	3.73
	0.5	20,968	3.01	2.92	3.16	3.54
7.5		22,337	3	2.91	3.16	3.57
	1.5	30,056	3.02	2.93	3.14	3.56
	0.5	19,591	3.04	2.93	3.36	3.61
10		20,960	3.03	2.92	3.36	3.61
	1.5	24,231	3.02	2.91	3.37	3.6

Table 4. *Cont.*

Model Dimension			Maximum Stress in Pipe		Maximum Stress in Patch	
Hole Diameter (mm)	Pipe Length (m)	Number of Elements	Von Mises Stress (MPa)	Hoop Stress (MPa)	Von Mises Stress (MPa)	Hoop Stress (MPa)
12.5	0.5	9,726	3	2.9	3.56	3.73
	1.5	11,095	2.99	2.89	3.55	3.72
		17,928	2.98	2.88	3.56	3.71
15	0.5	9,726	3	2.88	3.83	3.66
	1.5	11,095	2.99	2.87	3.82	3.65
		22,118	2.98	2.89	3.81	3.66
17.5	0.5	10,769	3.02	2.85	3.73	3.92
	1.5	12,138	3.01	2.84	3.73	3.92
		18,380	3.02	2.86	3.72	3.91
20.0	0.5	10,769	3.01	3.83	3.86	3.98
	1.5	12,138	3	3.82	3.85	3.98
		18,380	3.02	3.82	3.84	3.99

Additionally, to obtain accurate results, the postulated model was run based on different mesh conditions and number of elements. Considering the obtained solutions for the hoop and Von Mises stresses from the finite element solution it can be concluded that:

1. For the patch repaired pipe models and considering the same hole diameter it is clear that for models longer than 0.5 m, increasing the pipe length will not affect the maximum hoop and Von Mises stress values at the damage location in the pipe and also in the patch, significantly.

2. Increasing the hole diameter will increase the maximum value of hoop and Von Mises stresses in the patch. On the other hand increasing the hole diameter will not increase the maximum value of hoop and Von Mises stresses in the pipe significantly because of the reinforcing effect of the patch.

3. Comparing the results of various number of elements shows that if the distribution of the implemented finite elements is done properly, increasing the number of elements will not affect the results in pipe and patch significantly.

4. According to the results of finite element solution, seen in Figure 8, the maximum values of hoop and Von Mises stresses in the perforated pipe occur on the internal surface (inside) of the pipe and around the hole at the sides of the defect location.

Based on these results, the selected mesh type, finite elements, loading conditions, and applied boundary conditions in this section are appropriate for carrying out the next stages of the research and can be used to perform finite element modeling of the other three types of patch arrangements which will be discussed in the next section. Additionally, in order to obtain accurate results the value of 1.5 m is used for the pipe length in the next modeling stages.

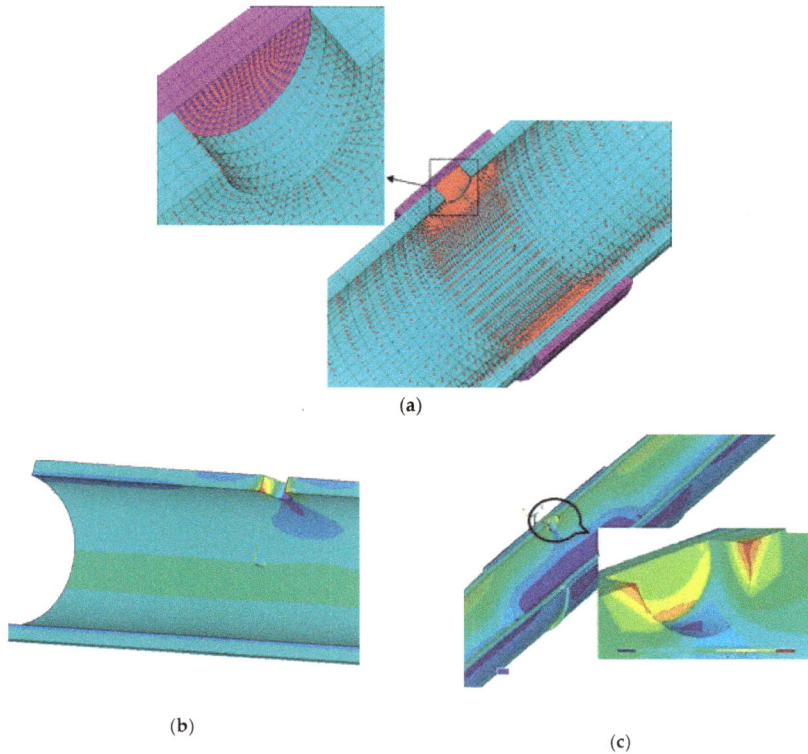

Figure 8. Finite element modeling of sectioned pipe and socket to show Von Mises stress distribution. (**a**) Mesh generation and the application of internal pressure. (**b**) Stress distribution in PE pipe with 1″ diameter circular hole. (**c**) Stress distribution in PE pipe and patch.

Comparing the Results of Different Patch Arrangements

Before investigating and discussing the complicated problem of buried patch repaired pipelines, for the better understanding, we decide to evaluate the efficiency of the patch repair in the reinforcement of the defected unburied pipe. Four different patch arrangements in the form of semi-cylindrical, circular-partial, square-partial and saddle fusion patches are modeled and discussed. The same procedure described in previous section is followed and used to calculate the maximum stresses in the patch and pipe loaded to an internal gas pressure of 4 bar. The number of finite elements used in this study to model unburied pipe and its patch for various patch arrangements are presented in Table 5.

The plotted curves in Figure 9 show the variation of maximum Von Mises stress in the defected patch repaired PE pipe for each kind of patch arrangement and also in unrepaired defected pipe, versus various hole diameters in the pipe. Additionally, the reduction percentage in maximum stress values resulted from the application of a special patch repair are shown and compared in Table 6.

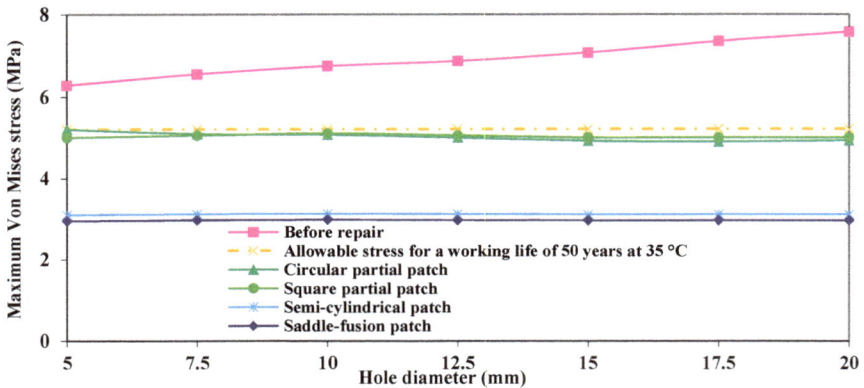

Figure 9. Curves for maximum Von Mises stresses in various patch arrangements for unburied pipe and patch arrangement.

Table 5. Number of finite elements used to model unburied pipe and its patch for various patch arrangements.

Hole Diameter (mm)	Number of Finite Elements			
	Saddle Fusion Patch	Semi-Cylindrical Patch	Circular Partial Patch	Square Partial Patch
5	18,889	18,044	8,198	6,823
7.5	22,.337	20,739	8,198	6,791
10	20,960	20,004	8,198	6,099
12.5	11,095	10,619	8,198	4,933
15	11,095	10,619	8,198	4,933
17.5	12,138	10,881	8,198	5,426
20	12,138	10,881	8,198	5,426

Table 6. Maximum Von Mises stress reduction percentage in the holed pipe for various patch arrangements under internal pressure.

Hole Diameter (mm)	Maximum Von Mises Stress Reduction in Pipe (%)			
	Square Partial Patch	Circular Partial Patch	Saddle Fusion Patch	Semi-Cylindrical Patch
5	12.5	7.9	48.6	46.5
7.5	17	15.9	50.9	48.9
10	20.4	20.5	53.2	51.1
12.5	23.4	25.1	55.4	53.5
15	26.3	29.7	57.1	54.8
17.5	29.9	32.3	59.2	57
20	33.5	34.2	61.4	59.3

It can be implied from Figure 9 and Table 6 that:

1. All four patch arrangements have significant effect on the reduction of the maximum Von Mises stresses in the reinforcement of the damaged section of the pipe.

2. The results of Table 6 show that the percentage of the maximum Von Mises stress reduction in the pipe increases for larger hole diameters.

3. The most effective patch in the reinforcement of the pipe is saddle fusion arrangement which reduces the maximum Von Mises stresses about 48%. For square-partial patch, although the effect is notable, it is less than the other 3 kinds of patches. This patch reduces the maximum Von Mises stresses by about 8%.

4. The aforementioned curves indicate that semi-cylindrical patch and saddle fusion patch show similar trends. Additionally, the results of circular-partial patch and square-partial patches are close to each other.

5. It can be implied that, the reinforcement effect of semi-cylindrical patch and saddle fusion patch are significantly more than the other two patch arrangements.

4.2.2. Finite Element Modeling and Solution for a Buried Patch Repaired Pipe with Circular Wall Holes

In this section we continue to investigate the studies in [9,10] with more details and under more critical loading conditions by superimposing various thermal loads in the form of temperature variations and the mechanical loads including surcharge loads, soil column weight, soil-pipe interaction and inside pressure of 4 bar applied to the repaired defected polyethylene pipe and its patch. The trench dimensions and pipe surroundings are selected based on Figure 10. Also, soil grades which are selected based on ASTM standards along with model dimensions are depicted on the computer simulation model in Figure 10.

The results of previous section show that the maximum Von Mises stress values are well below the allowable stresses (based on the design factor of 1.5) for working life of 50 years at 35 °C. Additionally lower values of safety factor (design factor) are applicable to design underground gas pipes. By selecting the value of 1.25 for safety factor, the values of allowable stresses for working life of 50 years at 35 °C will be 5.2 MPa and 6.5 MPa for PE80 pipe and PE100 patch materials respectively. We prefer to use these values for the following section of the research. In order to perform a proper finite element study on the underground structure, it is important to find an appropriate model dimension which the obtained results not to be dependent on the model size. Therefore, the finite element simulation of various model sizes are calculated and compared in Table 7.

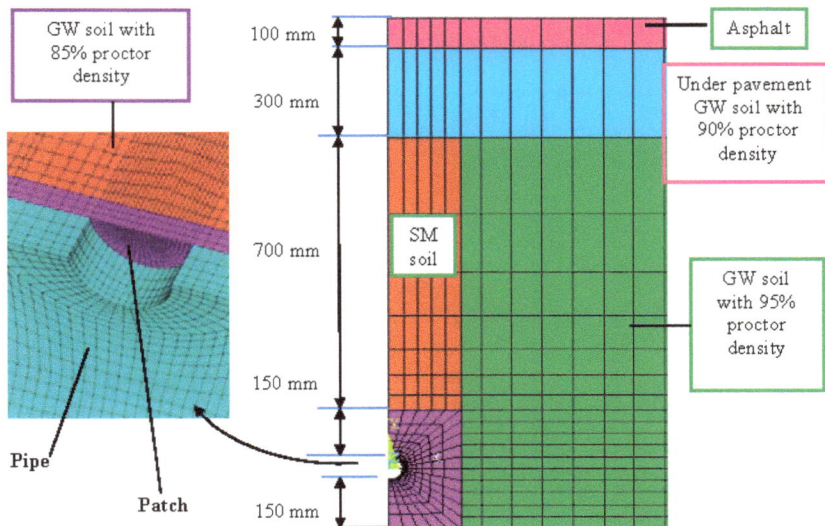

Figure 10. Finite element model of the patched pipe with surrounding soil subjected too internal pressure. GW: Well-graded gravels; SM: Silty sands [11].

The results of maximum Von Mises and hoop stresses in the pipe and its patch show that the upper surface dimension of $z \times x = (1.5 \text{ m}) \times (1.5 \text{ m})$ is appropriate to perform finite element modeling of this research, since larger models will not affect the stress values significantly.

Table 7. Simultaneous effect of soil column weight and internal pressure on maximum hoop and Von Mises stresses in defected pipe and patch for various x and z values.

Finite Element Model Specifications			Maximum Stress in Pipe		Maximum Stress in Patch	
Hole Diameter (mm)	Model Upper Surface Dimension $z \times x = A$	Number Finite Elements	Von Mises Stress (MPa)	Hoop Stress (MPa)	Von Mises Stress (MPa)	Hoop Stress (MPa)
	0.50 m × 0.4143 m	38,986	3.99	4.09	3.39	3.77
	0.50 m ×1.00 m	49,182	3.36	3.35	3.37	3.73
5	1.50 m × 1.00 m	57,594	3.35	3.34	3.37	3.73
	0.50 m × 1.50 m	73,537	3.25	3.22	3.36	3.71
	1.50 m × 1.50 m	81,068	3.24	3.22	3.35	3.7
	0.50 m × 0.4143 m	27,036	4.01	3.89	3.95	4.11
	0.50 m × 1.00 m	46,977	3.38	3.22	3.87	3.98
20	1.50 m × 1.00 m	62,011	3.37	3.22	3.86	3.98
	0.50 m × 1.50 m	74,326	3.28	3.11	3.84	3.94
	1.50 m × 1.50 m	82,994	3.27	3.1	3.38	3.95

The Simultaneous Effects of Thermo-Mechanical Loads on Semi-Cylindrical Patch Arrangement

Figure 11 shows the variations in the maximum values of Von Mises stresses in the buried PE80 pipe that is repaired by a 76.2 mm long semi-cylindrical patch arrangement, versus defect sizes in the form of circular holes under simultaneous effects of mechanical loads in the form of soil load, 4 bar internal pressure, vehicle wheel load and various thermal loads in the form of daily and seasonal temperature variations.

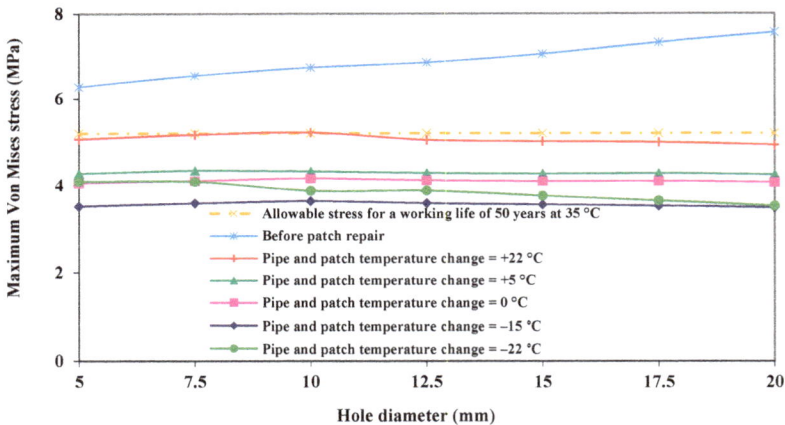

Figure 11. Maximum Von Mises stress variations in the buried pipe for various size circular hole defects repaired by semi-cylindrical patch (the simultaneous effect of soil load, internal pressure, vehicle wheel load and temperature variations).

The upmost curve shows the values of maximum Von Mises stress for the defected pipe before the application of patch repair and indicates that increasing hole diameter increases the maximum Von Mises stress significantly. The comparison of this curve with allowable stress value for a working life of 50 years (dashed line) shows that the imposed stresses in defected pipe are significantly higher than allowable stress values. For the other four curves which show the maximum Von Mises stress values in the defected polyethylene pipe repaired by semi-cylindrical patch, the stress values are well below the results of unrepaired defected pipe and also well below the allowable stress value. This means the aforementioned patch repair can strengthen the defected part of the pipe as well to transfer the gas fuel. Additionally, for the investigated defects at a constant temperature variation, the maximum Von Mises stresses remain approximately constant even by increasing the hole size. The curves for $-22\ ^\circ C$, $-15\ ^\circ C$, $0\ ^\circ C$ (no temperature changes), $+5\ ^\circ C$, and $+22\ ^\circ C$ temperature changes shows approximately similar trend with hole diameter increase where for these cases the minimum Von Mises stresses are not increasing significantly for larger hole sizes. Generally stating, the maximum Von Mises stresses in the pipe increases for higher temperature increases. That means the patch is more effective in reinforcing defected pipe for lower temperature changes. For example, for a fixed hole diameter of 20 mm, the stress reduction percentage in the pipe wall are 53%, 53%, 46%, 44%, and 35% for the temperature changes of $-22\,°C$, $-15\,°C$, $0\,°C$, $+5\,°C$, and $+22\,°C$ respectively. Also, the patch is more effective in reinforcing defected pipe for larger hole diameters. The maximum stress values are related to the seasonal temperature increase of $+22\,°C$.

Additionally, Figure 12 presents maximum values of Von Mises stresses in the 3" long semi-cylindrical patch arrangement versus pipe circular hole diameters under simultaneous effects of mechanical loads and various thermal loads in the form of daily and seasonal temperature variations. Based on the obtained results, the temperature variations have a significant effect on the maximum Von Mises stresses in the semi-cylindrical patch itself. The lowest curve on Figure 12 which shows the lowest values of maximum Von Mises stresses belongs to the situation where no temperature change is imposed to the pipe and patch at the burial depth under the ground that means it is the case with the minimum valves of the induced maximum Von Mises stresses compared to the other temperature changes. Additionally, the uppermost curve fits the data obtained for the maximum temperature change ($+22\,°C$ temperature increase, based on seasonal variations), which shows the case with highest values of induced maximum Von Mises stresses among others. For the semi-cylindrical PE100 patch material it can be implied that higher temperature changes (both temperature increase and temperature decrease) impose higher maximum Von Mises stresses. Also, for the cases of low temperature changes including $0\,°C$, $+5\,°C$, the maximum Von Mises stresses in the patch show an increase trend by increasing the pipe hole diameter, while for the cases of higher values of temperature changes including $-15\,°C$, $-22\,°C$, $+22\,°C$ the maximum Von Mises stresses will remain approximately constant even for larger hole diameters. Based on the calculated results, the semi-cylindrical patch can reinforce the proposed circular hole modeled defects efficiently. The only problem is that for higher temperature changes, the maximum Von Mises stress values in the patch itself can be critical which requires more research and investigation on the other patch configurations.

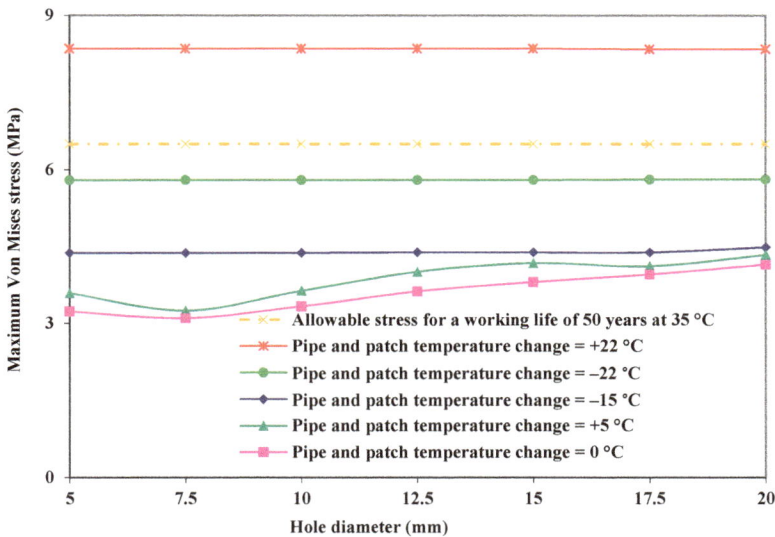

Figure 12. Maximum Von Mises stress in the semi-cylindrical patch for various sizes of circular hole defects (the simultaneous effect of mechanical and thermal loads).

The Simultaneous Effects of Thermo-Mechanical Loads on Circular Partial Patch Arrangement

In this section in order to more investigate to find a proper patch, a circular partial patch arrangement is designed and its reinforcing effect on the damaged underground PE gas pipe are being studied. The results of the FE simulation in the form of variations in the maximum values of Von Mises stresses in the buried PE80 pipe that is repaired by the circular partial patch arrangement, versus defect sizes in the form of circular hole under simultaneous effects of the previously mentioned mechanical loads and various thermal loads are depicted in Figure 13.

Similar to the previous case, as expected the upmost curve is related to the values of maximum Von Mises stresses for the defected pipe prior to circular-partial patch application and shows that for larger hole diameters, the maximum Von Mises stress increases significantly. Based on the results of the other five presented curves which shows the maximum Von Mises stresses in the defected polyethylene pipe repaired by circular-partial patch and comparing these curves with dashed line that indicates allowable stress values, the imposed stress values are well below the results of unrepaired defected pipe but slightly above the allowable stress values. Considering the case of 20 mm hole diameter, the stress reduction percentage in the pipe wall are 31%, 29%, 26%, 24%, and 22% for the temperature changes of −22 °C, −15 °C, 0 °C, +5 °C, and +22 °C respectively. The results imply that for the application of the aforementioned patch repair to reinforce the damaged part of the pipe some stress relief mechanisms must be employed too. Additionally, for the investigated defects at a constant temperature variation, the maximum on Mises stresses remain approximately constant even by increasing the hole size. The curves for −22 °C, −15 °C, 0 °C, +5 °C, and +22 °C temperature changes shows approximately similar trend with hole diameter increase. The maximum stress values are related to the seasonal temperature increase of +22 °C.

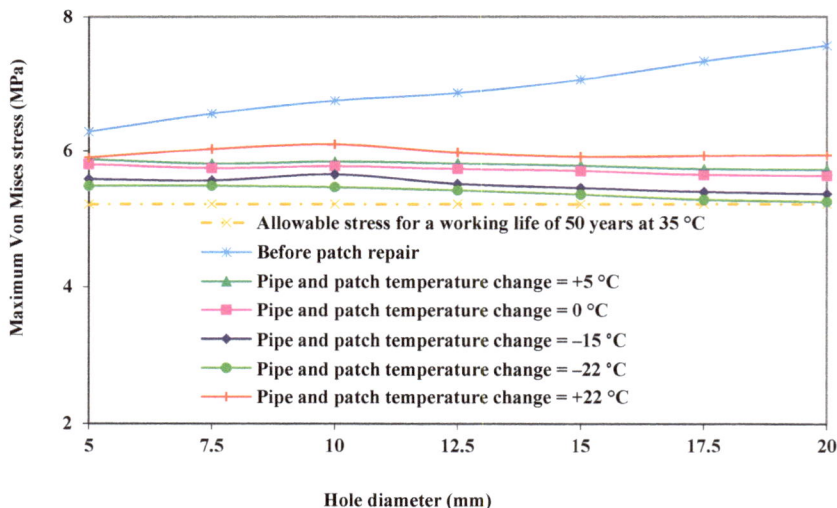

Figure 13. Maximum Von Mises stress variations in the buried pipe for various size circular hole defects repaired by circular partial patch (the simultaneous effect of soil load, internal pressure, vehicle wheel load and temperature variations).

Additionally, the maximum values of Von Mises stresses in the circular partial patch arrangement versus pipe circular hole diameters under simultaneous effects of aforementioned mechanical loads and various thermal loads in the form of daily and seasonal temperature variations are presented in Figure 14. According to the obtained results, the temperature variations have significant effect on the maximum Von Mises stress variations in the circular partial patch. The lowest curve on Figure 14 belongs to the situation where no temperature change is imposed to the pipe and patch at the burial depth under the ground that means it is the case with the minimum valves of the induced maximum Von Mises stresses. Additionally, the uppermost curve fits the data obtained for the maximum temperature change (+22 °C temperature decrease, based on seasonal variations), which shows the case with highest values of induced maximum Von Mises stresses among others. For the circular partial PE100 patch material it can be implied that higher temperature changes (both temperature increase and temperature decrease) impose higher maximum Von Mises stresses.

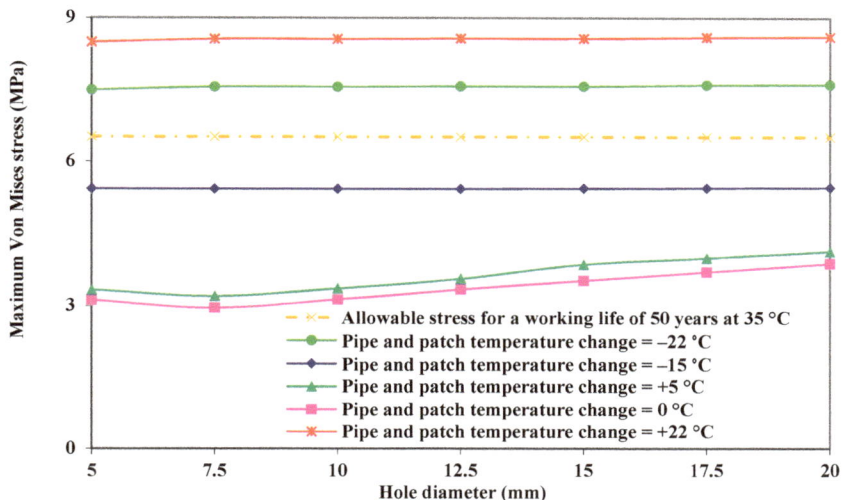

Figure 14. Maximum Von Mises stress in the circular partial patch for various sizes of circular hole defects (the simultaneous effect of mechanical and thermal loads).

For higher temperature changes, including a +22 °C temperature increase and a −22 °C temperature decrease, the maximum Von Mises stresses are above the allowable stress which means applying circular partial patches in these areas cannot be suggested. For lower temperature changes including −15 °C, 0 °C, and +5 °C the maximum Von Mises stresses are well below the allowable stress limit, which means the circular partial patch is applicable in the areas with the temperature changes up to the mentioned values.

The Simultaneous Effects of Thermo-Mechanical Loads on Square Partial Patch Arrangement

The same procedure which was discussed for semi-cylindrical and circular partial patches in the two previous sections is used to investigate square partial patch arrangement by finite element method. The results of Ansys simulation for the variations in the maximum values of Von Mises stresses in the buried PE80 pipe that is repaired by a square partial patch arrangement, versus defect sizes in the form of circular hole under simultaneous effects of mechanical loads in the form of soil load, 4 bar internal pressure, vehicle wheel load, and various thermal loads in the form of daily and seasonal temperature variations are depicted in Figure 15.

Comparing the curves showing the results of maximum Von Mises stress values for various temperature changes and the upmost curve which is related to the defected unrepaired pipe, shows that square-partial patch arrangement plays an important role in decreasing maximum stress values and strengthening the defected part of the pipe. For more understanding, considering the case of the 20 mm hole diameter, the results show the stress reduction percentage in the pipe wall are 29.5%, 29.2%, 25.6%, 24.3%, and 21.7% for the temperature changes of −22 °C, −15 °C, 0 °C, +5 °C, and +22 °C respectively, which shows that patch has more reinforcing effects for lower temperature changes. Even though, the square-partial patch has significant effect on reinforcing the defected pipe and decreasing the maximum Von Mises stress values, but comparing the results with the dashed line of allowable stress shows that the induced maximum Von Mises stresses are slightly higher than allowable stress values. Therefore, if we decide to use this kind of patch, more researches should be conducted to obtain some stress relief mechanisms. Additionally, the curves show similar trends for different temperature changes. Furthermore, comparing these curves with the results obtained for circular-partial patch shows approximately similar trends between these two cases.

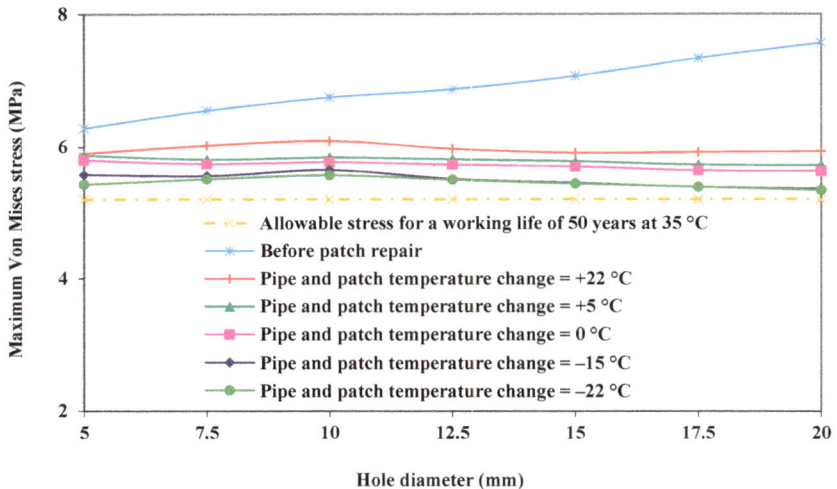

Figure 15. Maximum Von Mises stress variations in the buried pipe for various sizes of circular hole defects repaired by square partial patch (the simultaneous effect of soil load, internal pressure, vehicle wheel load and temperature variations).

Additionally, the maximum values of Von Mises stresses in the square partial patch arrangement versus pipe circular hole diameters under simultaneous effects of aforementioned mechanical loads and various thermal loads in the form of daily and seasonal temperature variations are presented in Figure 16. As can be implied, the temperature variations have a significant effect on the variation of the maximum Von Mises stresses in the square partial patch. The lowest curve on Figure 16 belongs to the situation where no temperature change is imposed to the pipe and patch at the burial depth under the ground that means it is the case with the minimum valves of the induced maximum Von Mises stresses. Additionally, uppermost curve fits the data obtained for the maximum temperature change (+22 °C temperature increase, based on seasonal variations), which shows the case with highest values of induced maximum Von Mises stresses among others. It is clear that for low temperature changes, including 0 °C and +5 °C, the maximum Von Mises stresses in the patch increases for larger hole diameters. For higher temperature changes including −15 °C, −22 °C, and +22 °C the results will remain approximately constant even by increasing hole diameter. For the square partial PE100 patch material it can be implied that higher temperature changes (both temperature increase and temperature decrease) impose higher maximum Von Mises stresses. The curves showing the data of patch stress results are below the allowable stress of the patch material except for the case of +22 °C temperature increase.

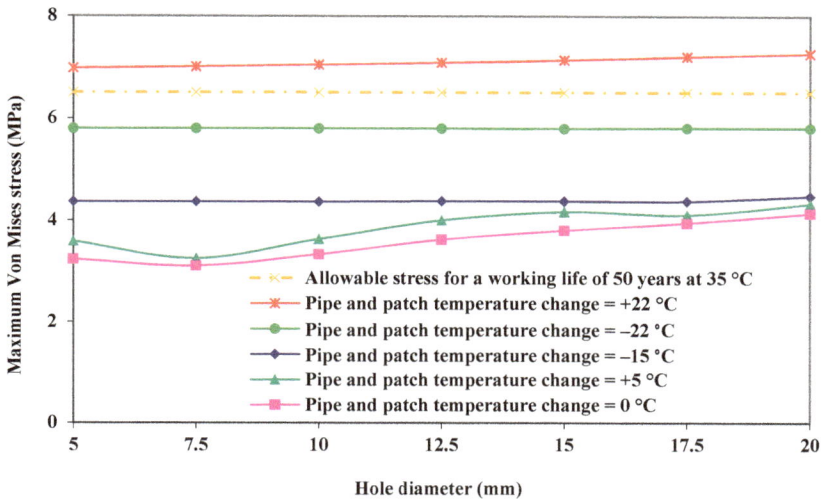

Figure 16. Maximum Von Mises stress in the square partial patch for various sizes of circular hole defects (the simultaneous effect of mechanical and thermal loads).

The Simultaneous Effects of Thermo-Mechanical Loads on Saddle Fusion Patch Arrangement

In the previous parts, three patch repair configurations were introduced and discussed in details. Finite element solutions for investigating the stress distribution in the aforementioned patch repairs to find the effectiveness of the proposed patch arrangements gave us the knowledge that semi-cylindrical patch configuration can effectively reinforce the defected part of the pipe. For the sake of finding a more reliable patch configuration we decide to investigate a full-cylindrical (called saddle fusion) patch repair. In order to verify the finite element model dimension, in Section 4.2.2 the variation of maximum Von Mises stresses in the repaired defected buried polyethylene gas pipe and its 3 inch long, saddle fusion patch under mechanical loads is well discussed. Figure 17 shows the variations in the maximum values of Von Mises stresses in the buried PE80 pipe that is repaired by a saddle fusion patch arrangement, versus defect sizes in the form of circular hole under simultaneous effects of mechanical loads in the form of soil load, 4 bar internal pressure, vehicle wheel load, and various thermal loads in the form of daily and seasonal temperature variations. The comparison of the curves resulted from saddle fusion patch repair for various temperature changes of −22 °C, −15 °C, 0 °C, +5 °C, and +22 °C with the upmost curve which shows the maximum Von Mises stress values for defected pipe before repair shows that the saddle fusion patch repair effectively reinforces the damaged part of the pipe to reliably transfer the gas. Additionally, comparing the mentioned five curves with the dashed line that indicates allowable stress values for PE80 pipe material shows that the maximum Von Mises stress values are well below the allowable stress. For more clarification considering the case of 20 mm hole diameter, the results show the stress reduction percentage in the pipe wall are 54%, 56%, 45%, 43%, and 36% for the temperature changes of −22 °C, −15 °C, 0 °C, +5 °C, and +22 °C respectively. All the five curves show similar trends.

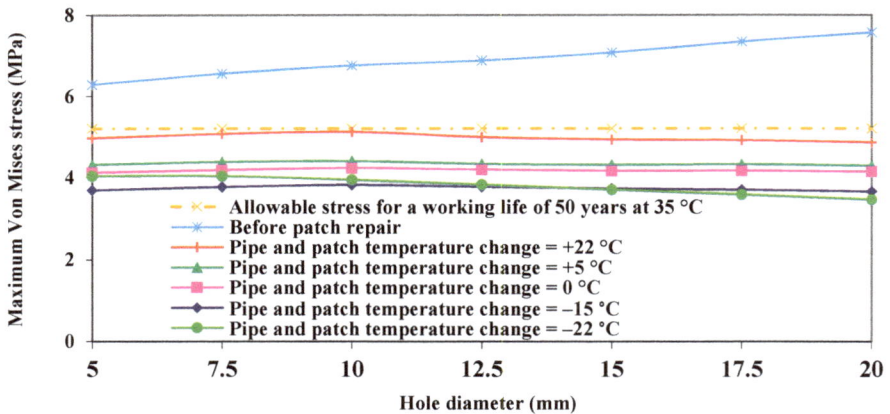

Figure 17. Maximum Von Mises stress in buried pipe for circular hole defects, saddle fusion patch repaired (simultaneous effects of soil load, internal pressure, vehicle load and temperature variations).

Figure 18 presents maximum values of Von Mises stresses in the 3 inch long, saddle fusion patch arrangement versus pipe circular hole diameters under simultaneous effects of previously mentioned mechanical and various discussed thermal loads. The obtained results show that, the temperature variations have significant effect on the maximum Von Mises stresses in the saddle fusion patch. similar to the 3 previously discussed patch arrangements the lowest curve on Figure 18 belongs to the situation where no temperature change is imposed to the pipe and patch at the burial depth under the ground that means it is the case with the minimum valves of the induced maximum Von Mises stresses. Additionally, the uppermost curve fits the data obtained for the maximum temperature change (+22 °C temperature increase, based on seasonal variations), which shows the case with highest values of induced maximum Von Mises stresses among others. For the saddle fusion PE100 patch material it can be implied that higher temperature changes (both temperature increase and temperature decrease) impose higher maximum Von Mises stresses. For all curves, the maximum Von Mises stress values in the patch will increase for larger hole diameters. For all cases, the maximum Von Mises stresses are well below the allowable stress limit for PE100 material which means the saddle fusion patch is well applicable to repair the proposed defects in even hot areas. Therefore, the results show the application of 3″ saddle fusion patch is advisable for the proposed loading condition.

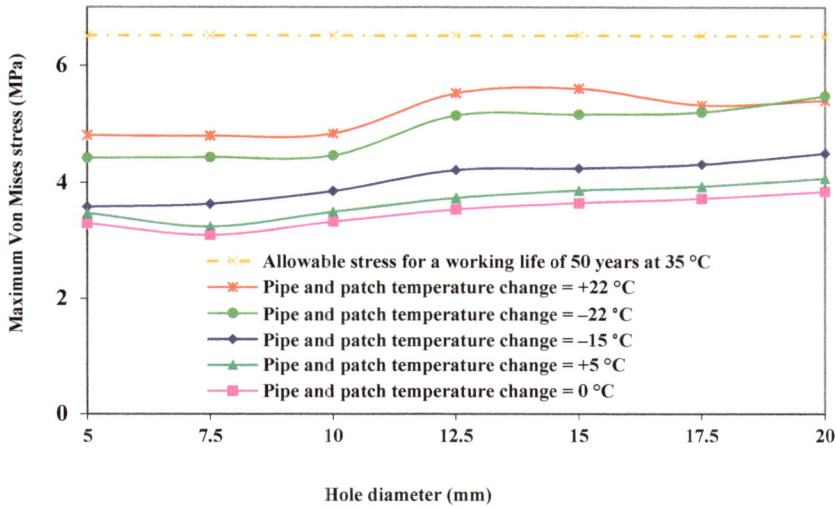

Figure 18. Maximum Von Mises stress in 3 inch long, saddle fusion patch for various sizes of circular hole defects (the simultaneous effect of mechanical and thermal loads).

5. Conclusions

In the present contribution a simple and effective computational method is developed to treat actual industrial fully three-dimensional complicated problems of large underground space. The finite element method is used to investigate stress distribution in buried gas pipe subjected to thermo-mechanical loads and stress concentrations due to changes in the geometry. This study was divided in two parts. In the first case, stress values were calculated for the socket joint of the buried PE pipe so that by applying the proper pipe joints, the stress be reduced to levels below the allowable values. For the second part, the effects of thermo-mechanical loads on the stress distribution in patch repaired buried pipes are well investigated. For this purpose, in this research, 3D finite element modeling of buried gas pipe and its socket/patch is performed using ANSYS software. Stress variations in the buried MDPE gas pipe and high density polyethylene (HDPE) socket/patch were fully investigated in a hot climate region to determine the critical stress values caused by stress concentrations at the pipe and socket/patch, enabling us to find the applicable method for joining or repairing MDPE gas transportation piping in such areas. The optimum burial depth was found to be 1.25 m, while the maximum and minimum ground surface temperatures at this depth were calculated to be 35 °C and 13 °C, respectively. Furthermore, the soil column weight above the pipe, the surcharge loads in terms of traffic load, the gas pressure of 4 bar inside the pipe and the stress concentrations due to a local change in geometry (in the form of damage) were imposed on the pipe and its socket/patch resulting in the following conclusions:

By considering the obtained plots for stress values in buried pipe and its socket and comparing it with allowable stress values for the pipe, the correct joining method can be investigated. Based on the results, maximum Von Mises stresses occurs at the middle of the socket internal surface while the maximum values of the aforementioned stresses in the socket occur where the internal surface of the socket joins the pipe outer surface. In both pipe and socket the maximum values of the aforementioned stresses are well below the allowable stresses and therefore the introduced socket joint can be used under the described working condition. Additionally, for the problem of the patch repaired buried pipes, the results show that all four patch configurations have significant reinforcing effect on the defected section of the buried pipe under the aforementioned thermo-mechanical loads. Meanwhile, the maximum Von Mises stresses in both pipe and saddle fusion patch are well below the allowable stress limit for PE material which means the saddle fusion patch is well applicable to repair the proposed defects in even hot areas. Therefore, the results show the application of introduced socket and also saddle fusion patch is advisable for the proposed loading condition.

Author Contributions: The two authors contributed to perform this research. R.K.-Z. conceived of the presented idea, developed the theory, created the ANSYS models, calculated mechanical stress on the models and discussed the results. P.A. gave feedback in all steps of the research work, made a critical review of the paper and also applied standards to use in this project.

Funding: This research received no external funding.

Acknowledgments: The first author highly appreciated prevalent help of his late friend Ali Nasirian for sharing his knowledge and experience in Finite Element modeling of underground space. His memory will be with me always.

Conflicts of Interest: The authors declare no conflict of interest.

Abbreviations

C	Design coefficient
D_i	Pipe inside diameter (mm)
D_o	Pipe outside diameter (mm)
E	Elastic modulus (MPa)
h	Pipe thickness (mm)
l	Pipe length in the model (mm)
MRS	Maximum reduced strength (MPa)
p_i	Pipe inside pressure (mm)
SDR	Standard dimension ratio (SDR = $\frac{D}{h}$)
t	Time (sec)
$T(y,t)$	Induced temperature in the soil (°C)
y	Depth of the pipe (cm or mm)
α	Thermal diffusivity (cm^2/s) at penetration depth
ΔT	Temperature change (°C)
ν	Poisson's ratio
σ	Normal stress in the pipe (MPa)
σ_e	Von Mises stress (MPa)
$\sigma_{\theta\theta}$	Tangential stress in the pipe (MPa)
σ_{rr}	Radial stress in the pipe (MPa)
σ_{zz}	Longitudinal stress in the pipe (MPa)

References

1. Khademi-Zahedi, R.; Makvandi, M.; Shishesaz, M. Technical analysis of the failures in a typical drilling mud pump during field operation. In Proceedings of the 22nd Annual International Conference on Mechanical Engineering—ISME2014, Ahvaz, Iran, 22–24 April 2014; Available online: https://www.civilica.com/Paper-ISME22-ISME22_347.html (accessed on 18 October 2018).
2. Khademi-Zahedi, R.; Makvandi, M.; Shishesaz, M. The applicability of casings and liners composite patch repair in Iranian gas and oil wells. In Proceedings of the 22nd Annual International Conference on Mechanical Engineering—ISME2014, Ahvaz, Iran, 22–24 April 2014; Available online: https://www.civilica.com/Paper-ISME22-ISME22_351.html (accessed on 18 October 2018).
3. Makvandi, M.; Bahmani, H.; Khademi-Zahedi, R. Technical analysis of the causes of blowout in BiBiHakimeh well No. 76. In Proceedings of the National Conference in New Research of Industry and Mechanical Engineering, Tehran, Iran, 17–18 September 2015; Available online: https://www.civilica.com/Paper-NRIME01-NRIME01_066.html (accessed on 18 October 2018).
4. Sampaio, J.H.B., Jr. Drilling Engineering Fundamentals. Master's Thesis, Curtin University Technology, Bentley, Australia, 2007.
5. *AASHTO Standard Specifications for Transportation Materials and Methods of Sampling and Testing*, 15th ed.; American Association of State Highway and Transportation Officials: Washington, DC, USA, 1990.

6. Moser, A.P.; Folkman, S. *Buried Pipe Design*; Book 2—Chapter 7; The McGraw-Hill Companies, Inc.: New York, NY, USA, 2001; 900p.

7. Goddard, J.B. *Plastic Pipe Design*; Technical Report 4.103; Advanced Drainage Systems, Inc.: Hilliard, OH, USA, 1994; pp. 1–30.

8. Khademi-Zahedi, R. Stress Distribution in Patch Repaired Polyethylene Gas Pipes. Master's Thesis, Shahid Chamran University, Ahvaz, Iran, 2011.

9. Khademi-Zahedi, R. Application of the finite element method for evaluating the stress distribution in burried damaged polyethylene gas pipes. *Undergr. Space* **2018**. [CrossRef]

10. Khademi-Zahedi, R.; Shishesaz, M. Application of a finite element method to stress distribution in buried patch repaired polyethylene gas pipe. *Undergr. Space* **2018**. [CrossRef]

11. Nasirian, A. Investigating the Application of Polyethylene Gas Pipes for Gas Transportation. Master's Thesis, Shahid Chamran University, Ahvaz, Iran, 2007.

12. Shishesaz, M.; Shishesaz, M.R. Applicability of medium density polyethylene Gas Pipes in Hot Climate Areas of South-West Iran. *Iran. Polym. J.* **2008**, *17*, 503–517.

13. Khademi-Zahedi, R.; Shishesaz, M.; Karamy-Moghadam, A.; Veisi-Ara, A. Stress analysis in defected buried polyethylene pipes for gas distribution networks. In Proceedings of the National Conference in New Research of Industry and Mechanical Engineering, Tehran, Iran, 17–18 September 2015.

14. Plastic Pipe and Building Products. In *ASTM Annual Book of ASTM Standards*; American Society for Testing and Material: Philadelphia, PA, USA, 1991.

15. The Plastics Pipe Institute. *Handbook of Polyethylene Pipe*; Plastics Pipe Institute, Inc.: Washington, DC, USA, 2006; Chapter 6; pp. 157–260.

16. Corrugated Polyethylene Pipe Association. *Structural Design Method for Corrugated Polyethylene Pipe*; Corrugated Polyethylene Pipe Association: Washington, DC, USA, 2000.

17. European Standard. EN1555-2, Plastic Piping Systems for the Supply of Gaseous Fuels-Polyethylene (PE)-Part 2: Pipes, European Committee for Standardization. Available online: https://webstore.ansi.org/RecordDetail.aspx?sku=DIN+EN+1555-2%3A2010 (accessed on 18 October 2018).

18. Plastic Pipe Institute. *Polyethylene Gas Pipes Systems, Uponor Aldyl Company Installation Guide*; Plastics Pipe Institute: Washington, DC, USA, 2004.

19. Egan, B.; Qenos, A.H. PE100, and Its Advantages When Used for Ploughing & Trenchless Technology. Available online: www.masterplumbers.com/plumbwatch/pipes99/pe100.doc (accessed on 18 October 2018).

20. Tabatabaii, A.; Ameri, M.; Behbahani, H. Prediction of Asphalt Cover Temperature in City of Ahvaz (Iran). *J. Eng.* **1999**, *7*, 22. (In Persian)

21. International Organization for Standardization. ISO12162, Thermo plastics Materials for Pipes and Fittings for Pressure Applications–Classification and Designation–Overall Service (Design) Coefficient. Available online: https://www.iso.org/standard/43865.html (accessed on 18 October 2018).

22. Amiri, F.; Milan, D.; Shen, Y.; Rabczuk, T.; Arroyo, M. Phase-field modeling of fracture in linear thin shells. *Theor. Appl. Fract. Mech.* **2014**, *69*, 102–109. [CrossRef]

23. Areias, P.; Rabczuk, T. Finite strain fracture of plates and shells with configurational forces and edge rotation. *Int. J. Numer. Methods Eng.* **2013**, *94*, 1099–1122. [CrossRef]

24. Areias, P.; Rabczuk, T. Steiner-point free edge cutting of tetrahedral meshes with applications in fracture. *Finite Elem. Anal. Des.* **2017**, *132*, 27–41. [CrossRef]

25. Areias, P.; Msekh, M.A.; Rabczuk, T. Damage and fracture algorithm using the screened Poisson equation and local remeshing. *Eng. Fract. Mech.* **2016**, *158*, 116–143. [CrossRef]

26. Areias, P.; Rabczuk, T.; Msekh, M. Phase-field analysis of finite-strain plates and shells including element subdivision. *Comput. Methods Appl. Mech. Eng.* **2016**, *312*, 322–350. [CrossRef]

27. Areias, P.M.A.; Rabczuk, T.; Camanho, P.P. Finite strain fracture of 2D problems with injected anisotropic softening elements. *Theor. Appl. Fract. Mech.* **2014**, *72*, 50–63. [CrossRef]

28. Chau-Dinh, T.; Zi, G.; Lee, P.S.; Song, J.H.; Rabczuk, T. Phantom-node Method for Shell Models with Arbitrary Cracks. *Comput. Struct.* **2012**, *92–93*, 242–256. [CrossRef]

29. Ghorashi, S.; Valizadeh, N.; Mohammadi, S.; Rabczuk, T. T-spline based XIGA for Fracture Analysis of Orthotropic Media. *Comput. Struct.* **2015**, *147*, 138–146. [CrossRef]

30. Hamdia, K.; Zhuang, X.; Silani, M.; He, P.; Rabczuk, T. Stochastic analysis of the fracture toughness of polymeric nanoparticle composites using polynomial chaos expansions. *Int. J. Fract.* **2017**, *206*, 215–227. [CrossRef]

31. KhademiZahedi, R.; Alimouri, P.; Nguyen-Xuan, H.; Rabczuk, T. Crack detection in a beam on elastic foundation using differential quadrature method and the Bees algorithm optimization. *Proc. Int. Conf. Adv. Comput. Mech.* **2017**, *36*, 439–460.

32. Nguyen-Thanh, N.; Kiendl, J.; Nguyen-Xuan, H.; Wuchner, R.; Bletzinger, K.U.; Bazilevs, Y.; Rabczuk, T. Rotation free isogeometric thin shell analysis using PHT-splines. *Comput. Methods Appl. Mech. Eng.* **2011**, *200*, 3410–3424. [CrossRef]

33. Nguyen-Thanh, N.; Valizadeh, N.; Nguyen, M.N.; Nguyen-Xuan, H.; Zhuang, X.; Areias, P.; Zi, G.; Bazilevs, Y.; De Lorenzis, L.; Rabczuk, T. An extended isogeometric thin shell analysis based on Kirchhoff-Love theory. *Comput. Methods Appl. Mech. Eng.* **2015**, *284*, 265–291. [CrossRef]

34. Nguyen-Thanh, N.; Zhou, K.; Zhuang, X.; Areias, P.; Nguyen-Xuan, H.; Bazilevs, Y.; Rabczuk, T. Isogeometric analysis of large-deformation thin shells using RHT-splines for multiple-patch coupling. *Comput. Methods Appl. Mech. Eng.* **2017**, *316*, 1157–1178. [CrossRef]

35. Rabczuk, T. Computational Methods for Fracture in Brittle and Quasi-Brittle Solids: State-of-the-art Review and Future Perspectives. *ISRN Appl. Math.* **2013**, *2013*, 849231. [CrossRef]

36. Rabczuk, T.; Areias, P.M.A.; Belytschko, T. A meshfree thin shell method for non- linear dynamic fracture. *Int. J. Numer. Methods Eng.* **2007**, *72*, 524–548. [CrossRef]

37. Rabczuk, T.; Gracie, R.; Song, J.H.; Belytschko, T. Immersed particle method for fluid-structure interaction. *Int. J. Numer. Methods Eng.* **2010**, *81*, 48–71. [CrossRef]

38. Ren, H.; Zhuang, X.; Cai, Y.; Rabczuk, T. Dual-Horizon Peridynamics. *Int. J. Numer. Methods Eng.* **2016**, *108*, 1451–1476. [CrossRef]

39. Ren, H.; Zhuang, X.; Rabczuk, T. Dual-horizon peridynamics: A stable solution to varying horizons. *Comput. Methods Appl. Mech. Eng.* **2017**, *318*, 762–782. [CrossRef]

40. KhademiZahedi, R.; Alimouri, P. Finite element model updating of a large structure using multi-setup stochastic subspace identification method and bees optimization algorithm. *Front. Struct. Civ. Eng.* **2018**, in press.

41. J-M Manufacturing Company Inc. International MDPE Yellow Gas Pipe. Available online: http://www.jmeagle.com/sites/default/files/Products_MDPE%20Yellow%20Gas.pdf (accessed on 6 December 2017).

![energies logo] *energies*

MDPI

Article

Characterizing Flexoelectricity in Composite Material Using the Element-Free Galerkin Method

Bo He [3,4], Brahmanandam Javvaji [4] and Xiaoying Zhuang [1,2,*]

1   Division of Computational Mechanics, Ton Duc Thang University, 700000 Ho Chi Minh City, Viet Nam
2   Faculty of Civil Engineering, Ton Duc Thang University, 700000 Ho Chi Minh City, Viet Nam
3   College of Civil Engineering, Tongji University, 1239 Siping Road, Shanghai 200092, China;
    bohe@ikm.uni-hannover.de
4   Institute of Continuum Mechanics, Leibniz Universität Hannover, Appelstr. 11, 30167 Hannover, Germany;
    brahmanandam@ikm.uni-hannover.de
*   Correspondence: xiaoying.zhuang@tdtu.edu.vn

Received: 30 October 2018; Accepted: 11 December 2018; Published: 16 January 2019

Abstract: This study employs the Element-Free Galerkin method (EFG) to characterize flexoelectricity in a composite material. The presence of the strain gradient term in the Partial Differential Equations (PDEs) requires C^1 continuity to describe the electromechanical coupling. The use of quartic weight functions in the developed model fulfills this prerequisite. We report the generation of electric polarization in a non-piezoelectric composite material through the inclusion-induced strain gradient field. The level set technique associated with the model supervises the weak discontinuity between the inclusion and matrix. The increased area ratio between the inclusion and matrix is found to improve the conversion of mechanical energy to electrical energy. The electromechanical coupling is enhanced when using softer materials for the embedding inclusions.

Keywords: flexoelectricity; meshless method; composite; size effect; level set technique

---

1. Introduction

An energy harvester utilizing the electromechanical coupling effect has been applied in various applications, ranging from sensors [1–3] to biomedical devices [4,5] at both micro- and nano-scales [6–8]. The well-known electromechanical coupling effect, piezoelectricity, generates the electrical polarization under mechanical deformation only in the non-centrosymmetric material. Flexoelectricity is another type of electromechanical coupling, which describes the coupling between the electrical potential and strain gradient. Unlike piezoelectricity, flexoelectricity can exist in any dielectric material regardless of the material's inner structure. In the meantime, experimental studies [9,10] observed an unexpected giant flexoelectric effect in barium strontium titanate, which suggested that at the micro-/nano-scale, the flexoelectric effect outperforms the piezoelectric effect. These superior properties of flexoelectricity at a small scale have drawn extensive research attention towards the fundamental investigation of this electromechanical coupling from the molecular scale to the continuum scale. Atomic-level density functional theories [11–13] and molecular dynamics simulations [14,15] explored the mechanism of inducing flexoelectric polarization under deformation. However, continuum scale studies can lead to better understanding of the practical applicability of the flexoelectric property in designing sensors and actuators.

The characterization of flexoelectricity with existing numerical methods, like the Finite Element Method (FEM), requires remarkable modifications to handle the strain gradient term. Zhang et al. [16] used conventional FEM to evaluate the piezo- and flexo-electric effects by solving the partial differential equations for the displacement and electric potential in a decoupled manner. The conventional FEM works on the basis of C^0 continuity and is uncertain when higher order strain terms are present (challenging to hold C^1 continuity). Nanthakumar et al. [17,18] used the Mixed Finite Element Formulation (MFEF) to maximize the flexoelectric effect through topology optimization of barium titanate material. However, the MFEF method suffers from high computational cost [19,20]. Isogeometric Analysis (IGA) is another numerical method, which ensures the required C^1 continuity by employing the higher-order shape functions [21–23]; whereas, IGA strongly depends on the geometrical symmetry conditions to reduce the computational cost [24,25]. Abdollahi et al. [26,27] developed a meshless method to study the flexoelectric response of dielectric material both in cantilever beam and truncated pyramid shapes. Their study further suggested that the simplified analytical model is unable to capture the flexoelectricity in multi-dimensional geometries. These studies have used structures with a ceramic material such as barium titanate or strontium titanate. Applications of these materials have a limited scope due to the small operational strain and high-stress conditions [28].

Experimental studies revealed that composites made up of two or many materials are promising to support high stress operating conditions. For example, nano-indentation tests of a bilayer cantilever beam structure have shown good mechanical properties [29,30]. A study on the polymer-based composite with graphene oxide has shown enhanced piezoelectric performance with greater flexibility [31]. Synthesized polymeric composite with piezoelectric zirconia titanate material promised to exhibit good energy-harvesting solutions [32]. However, these findings critically depend on the piezoelectric properties of the material. On the other hand, it is possible to create the piezoelectric composites without using a material with piezoelectric properties. The flexoelectric fiber-reinforced composite [33,34] and multi-material-based flexoelectric composites [35] are some example studies in this direction. The flexoelectric composites develop large strain gradients near the material discontinuity when uniformly deformed, which generates the electrical polarization through flexoelectricity. However, the numerical modeling other than conventional FEM requires extra care when dealing with the material and geometry discontinuity between the composite constituents. Lagrange multipliers [36,37] and the global enrichment approach [38,39] are the few techniques to treat the material discontinuity. The applicability of these methods is cumbersome when using complex geometries in 2D structures [40,41]. The Extended Finite Element Method (XFEM) has been used for the level set technique to describe the weak discontinuity between the inclusion and matrix composite [42–44]. In addition, the level set technique also allows using the multiple randomly-distributed inclusions inside the matrix.

The current work employs the Element-Free Galerkin (EFG) method [45–49] along with the level set technique to characterize flexoelectricity in a composite. It is assumed that the composite is a combination of embedding matrix material and a non-piezoelectric material (inclusion). The possible generation of electrical voltage due to the compressive loading of the composite is investigated. This study is extended to find the dependence of the electrical voltage on the concentration of inclusions in the embedding matrix. The numerical model is validated with standard examples and then utilized to explore the flexoelectricity in a composite. Section 2 composes the details of the EFG. Numerical validation and example case studies are presented in Section 3. Section 4 concludes the work.

2. Simulation Method

The enthalpy density $\mathcal{H}$ for a dielectric solid with the piezoelectric and flexoelectric effect is [50–52]:

$$\mathcal{H}(\epsilon_{ij}, E_i, \epsilon_{jk,l}, E_{i,j}) = \frac{1}{2}\mathbb{C}_{ijkl}\epsilon_{ij}\epsilon_{kl} - e_{ikl}E_i\epsilon_{kl} - \mu_{ijkl}E_i\epsilon_{jk,l} - \frac{1}{2}\kappa_{ij}E_iE_j \,, \tag{1}$$

where $E_i = -\phi_{,i}$ is the electric field; ϕ_i being the electric potential in the i direction; ϵ_{ij} is the mechanical strain; $\mathbb{C}_{ijkl}$ is the fourth-order elastic moduli tensor; e_{ikl} is the third-order tensor of piezoelectricity; μ_{ijkl} is the fourth-order flexoelectric tensor; and κ_{ij} is the second-order dielectric tensor. Appendix A covers further details about the weak form of Equation (1) and possible boundary conditions. The resulting governing equations are solved for unknown variables displacement u and electric potential ϕ using Moving Least Squares approximation (MLS). Belytschko et al. [53–55] introduced the MLS approximation within the EFG. According to the MLS, the unknowns (u and ϕ) for the center node are approximated by the neighboring nodes in its support domain (Figure 1a). If the support domain encounters a material discontinuity, corresponding support nodes are enriched. The weak discontinuity across the material interface in the composite is modeled with the level set function with absolute sign distance function $\Psi(x)$. Figure 1b shows the schematic illustration of the level set function and its first-order-derivative across the interface. The local approximation of displacement u and electric potential ϕ at location x are given as:

$$u^h(x) = \sum_{I=1}^{N} \Phi_I(x) u_I + \sum_{J=J}^{M} \Phi_J(x) \Psi(x) u_J;$$

$$\phi^h(x) = \sum_{I=1}^{N} \Phi_I(x) \phi_I + \sum_{J=J}^{M} \Phi_J(x) \Psi(x) \phi_J,$$

(2)

where $\Phi(x)$ is the shape function at position x, u and ϕ are the nodal displacements and electric potential, and N and M are the number of support and enriched nodes, respectively. More details about $\Phi(x)$ are given in Appendix B. Hereafter, we write the approximation in a simplified form $u^h = \Phi_u u^{std} + \Phi_u \Psi u^{enr}$. The derivatives for unknowns u and ϕ (Equation (2)) contribute by both support nodes (u^{std}, ϕ^{std}) and enriched nodes (u^{enr}, ϕ^{enr}), which are given as:

$$\frac{\partial u^h}{\partial x} = \frac{\partial \Phi_u}{\partial x} u^{std} + \frac{\partial \Phi_u}{\partial x} \Psi u^{enr} + \Phi_u \frac{\partial \Psi}{\partial x} u^{enr};$$

$$\frac{\partial^2 u^h}{\partial x^2} = \frac{\partial^2 \Phi_u}{\partial x^2} u^{std} + \frac{\partial^2 \Phi_u}{\partial x^2} \Psi u^{enr} + \Phi_u \frac{\partial^2 \Psi}{\partial x^2} u^{enr} + 2 \frac{\partial \Phi_u}{\partial x} \frac{\partial \Psi}{\partial x} u^{enr};$$

$$\frac{\partial \phi^h}{\partial x} = \frac{\partial \Phi_\phi}{\partial x} \phi^{std} + \frac{\partial \Phi_\phi}{\partial x} \Psi \phi^{enr} + \Phi_\phi \frac{\partial \Psi}{\partial x} \phi^{enr}.$$

(3)

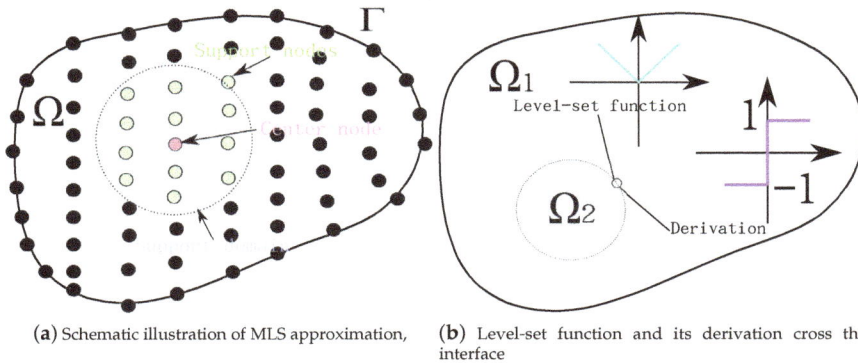

(**a**) Schematic illustration of MLS approximation,   (**b**) Level-set function and its derivation cross the interface

Figure 1. (a) Schematic illustration of Moving Least Squares (MLS) approximation and (b) the level-set function and its derivation across the interface.

The discrete equilibrium equations involving Equation (3) are given as:

$$\begin{bmatrix} A_{uu} & A_{u\phi} \\ A_{u\phi}^T & A_{\phi\phi} \end{bmatrix} \cdot \begin{bmatrix} u \\ \phi \end{bmatrix} = \begin{bmatrix} f_u \\ f_\phi \end{bmatrix},$$

(4)

where:

$$A_{uu} = \sum_e \int_{\Omega_e} (B_u + B_u^{enr}) \mathbb{C} (B_u + B_u^{enr})^T d\Omega_e;$$

(5)

$$A_{u\phi} = \sum_e \int_{\Omega_e} \left((\boldsymbol{B}_u + \boldsymbol{B}_u^{enr})\boldsymbol{e}(\boldsymbol{B}_\phi + \boldsymbol{B}_\phi^{enr})^T + (\boldsymbol{H}_u + \boldsymbol{H}_u^{enr})\boldsymbol{\mu}^T(\boldsymbol{B}_\phi + \boldsymbol{B}_\phi^{enr})^T \right) d\Omega_e; \tag{6}$$

$$A_{\phi\phi} = -\sum_e \int_{\Omega_e} (\boldsymbol{B}_\phi + \boldsymbol{B}_\phi^{enr})\boldsymbol{\kappa}(\boldsymbol{B}_\phi + \boldsymbol{B}_\phi^{enr})^T d\Omega_e; \tag{7}$$

$$f_u = \sum_e \int_{\Gamma_{te}} \boldsymbol{N}_u^T \boldsymbol{t}_\Gamma d\Gamma_{te}; \tag{8}$$

$$f_\phi = -\sum_e \int_{\Gamma_{De}} \boldsymbol{N}_\phi^T w d\Gamma_{De}; \tag{9}$$

where e refers to element number. Further details about the individual $\boldsymbol{B}$ and $\boldsymbol{H}$ matrices are included in Appendix C. The Lagrange multiplier method is used for imposing mechanical and electrical boundary conditions. In this study, the plane strain condition is assumed. Material property matrices $\mathbb{C}$, $\boldsymbol{\kappa}$, $\boldsymbol{e}$, $\boldsymbol{\mu}$ are given as:

$$\mathbb{C} = \frac{E}{(1+\nu)(1-2\nu)} \begin{bmatrix} 1-\nu & \nu & 0 \\ \nu & 1-\nu & 0 \\ 0 & 0 & (\frac{1}{2}-\nu) \end{bmatrix}, \tag{10}$$

$$\boldsymbol{\kappa} = \begin{bmatrix} k_{11} & 0 \\ 0 & k_{33} \end{bmatrix}, \tag{11}$$

$$\boldsymbol{e}^T = \begin{bmatrix} 0 & 0 & e_{15} \\ e_{31} & e_{33} & 0 \end{bmatrix}, \tag{12}$$

$$\boldsymbol{\mu} = \begin{bmatrix} \mu_{11} & \mu_{12} & 0 & 0 & 0 & \mu_{44} \\ 0 & 0 & \mu_{44} & \mu_{12} & \mu_{11} & 0 \end{bmatrix}. \tag{13}$$

3. Numerical Examples

In this section, we validate the numerical model with benchmark problems: electromechanical characteristics of a cantilever beam under mechanical and electrical loading conditions and coupling effects in a mechanically-compressed truncated pyramid. Note that these problems are free from the material discontinuity. The obtained results are validated with the reported literature. The complete numerical model with material discontinuity and local enrichment is employed to estimate the characteristics of the composite material.

3.1. Cantilever Beam

Figure 2a,b shows the simulation setup for open circuit mechanical loading and close circuit electrical loading, respectively. The length to thickness ratio L/h for the cantilever beam is 20. Table 1 reports the various material parameters [21].

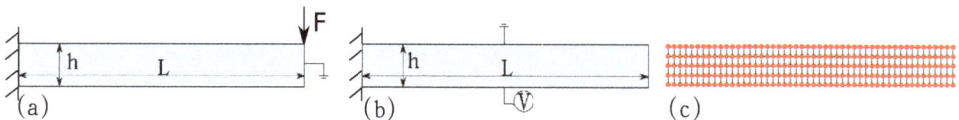

Figure 2. Schematic illustration of the cantilever beam under (**a**) mechanical loading and (**b**) electrical loading. (**c**) MLS discretization with the red solid point representing the nodes.

Table 1. Material properties.

Name	Symbol	Value
Poisson's ratio	ν	0.37
Young's modulus	E	100 GPa
Piezoelectric constant	e_{31}	$-4.4\,\text{nC/m}^2$
Flexoelectric constant	μ_{12}	$1\,\mu\text{C/m}$
Dielectric constant	$\kappa_{11}; \kappa_{33}$	11 nC/Vm; 12.48 nC/Vm
Electric susceptibility	χ	1408

3.1.1. Mechanical Loading

A point load $F = 100\,\mu\text{N}$ is applied on the upper right edge of the cantilever beam (Figure 2a), and the electric potential is constrained to zero. Electromechanical coupling induces the electrical energy under point load deformation. The conversion from mechanical to electrical energy (k_{eff}^2) is defined as:

$$k_{eff}^2 = \frac{W_{elec}}{W_{mech}} = \frac{\int E \cdot \kappa \cdot E}{\int \epsilon : \mathbb{C} : \epsilon}. \tag{14}$$

The present model is simplified by assuming that the transversal piezoelectric ($e = e_{31}$) and flexoelectric ($\mu = \mu_{12}$) components are the only non-zero in Equations (12) and (13). The Poisson effect is also ignored. The results of this simplified model are compared with the analytical model derived by Majdoub et al. [50]. The analytical solution for k_{eff} is:

$$k_{eff} = \frac{\chi}{1 + \chi} \sqrt{\frac{\kappa}{Y}\left(e^2 + 12\left(\frac{\mu}{h}\right)^2\right)}, \tag{15}$$

where the normalized piezoelectric constant is:

$$e' = \frac{k_{eff}}{k_{piezo}}, \tag{16}$$

where k_{piezo} is obtained by neglecting flexoelectricity coefficient μ in Equation (15).

Figure 3a plots the comparison between the present model and the analytical solution, where $h' = -eh/\mu$ is the normalized beam thickness for the open circuit mechanical loading condition. The variation between e' and h' from the present model agrees with the analytical solution from Equation (16). This proves that the present model correctly estimates the electromechanical coupling in a non-piezoelectric beam under bending deformation. The coupling between mechanical and electric energy is strongly mediated by the flexoelectricity due to the low piezoelectric constant of the material (refer to Table 1). Figure 3b shows the distribution of the generated electric potential under the flexoelectric effect.

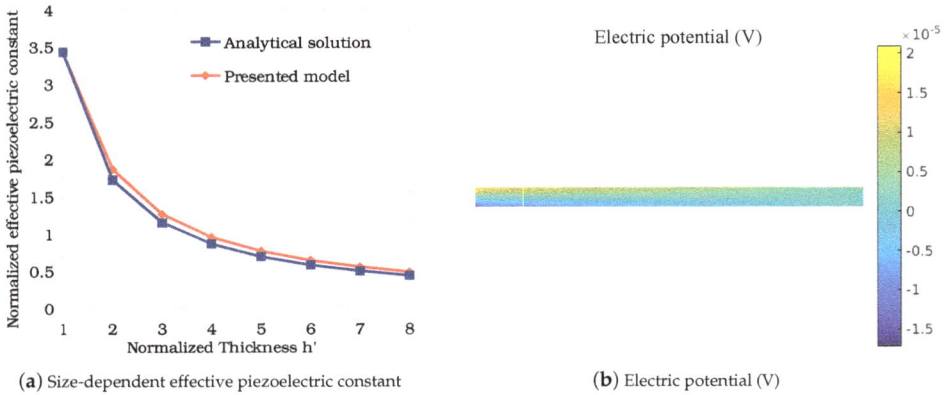

(**a**) Size-dependent effective piezoelectric constant   (**b**) Electric potential (V)

Figure 3. Calculation results of the (**a**) size-dependent effective piezoelectric constant and (**b**) the electric potential profile for a fully-coupled cantilever beam.

3.1.2. Electrical Loading

In this section, we study a cantilever beam (Figure 2b) under pure electric loading, which served as an actuator. The bottom edge of the cantilever beam (Figure 2b) enforced an electrical loading of -20V and grounded the top edge. There is no external mechanical loading in the cantilever beam. Figure 4 displays the displacement and electric potential profiles. The electric field across the beam in the y-direction is plotted in Figure 5. The electric field gradients near the top (negative sign) and bottom (positive sign) surface of the cantilever beam are due to the effect of converse flexoelectricity. This phenomenon generates mechanical stress at the top and bottom surface and thus deforms the beam. The displacement and electric potential distribution on the cantilever beam are in good agreement with earlier findings based on IGA analysis [21].

(**a**) Displacement in the x and y directions (m)

(**b**) Electric potential (V)

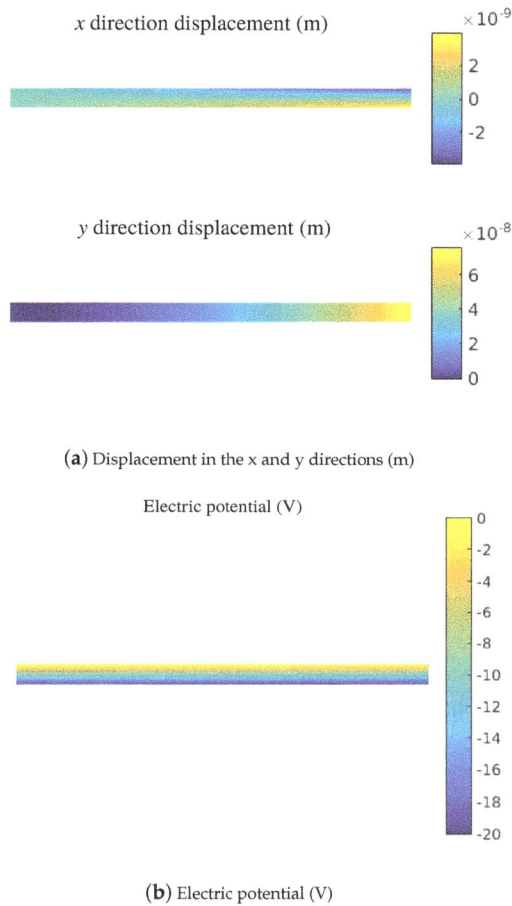

Figure 4. Calculation result of the cantilever beam under electric loading: (**a**) beam displacement; (**b**) electric potential profile.

3.2. Compress Truncated Pyramid

In this section, we have extended the validation for the compression of a truncated pyramid (Figure 6). Due to its different top (a_1) and bottom (a_2) edge lengths, the applied uniform force F generates different tractions at the top and bottom edges, which results in a longitudinal strain gradient. The top edge of the pyramid is grounded, and a uniform force of $-6e6$ N is applied to it. The aspect ratio h of the pyramid is set as 75 μm, and the bottom edge length is 225 μm. The remaining material parameters are listed in Table 1. Figure 7 shows the developed strain in the y-direction and the electromechanical coupling-induced electric potential profiles. The numerical values and the field distributions are in good agreement with earlier reports [21]. This represents that the present model accurately estimated the electromechanical coupling.

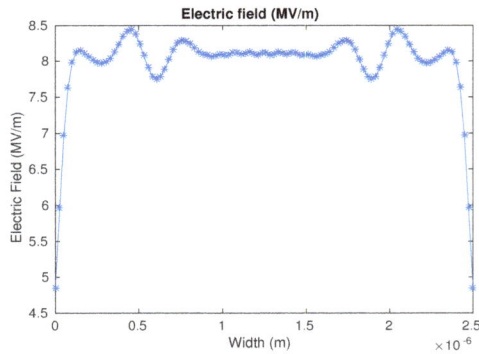

Figure 5. Calculation result of the electric field profile across the beam in the y direction at the mid-length of the beam.

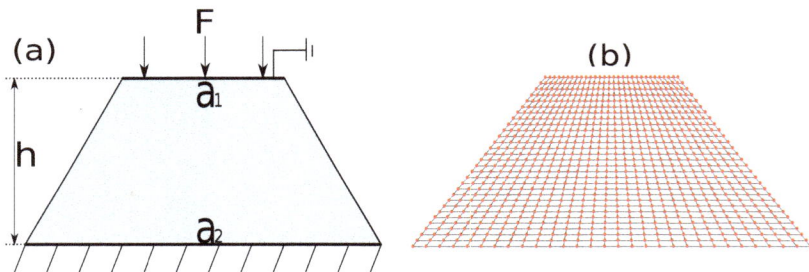

Figure 6. Schematic illustration of: (**a**) the pyramid case and its boundary condition; (**b**) MLS discretization with the red solid point representing the nodes.

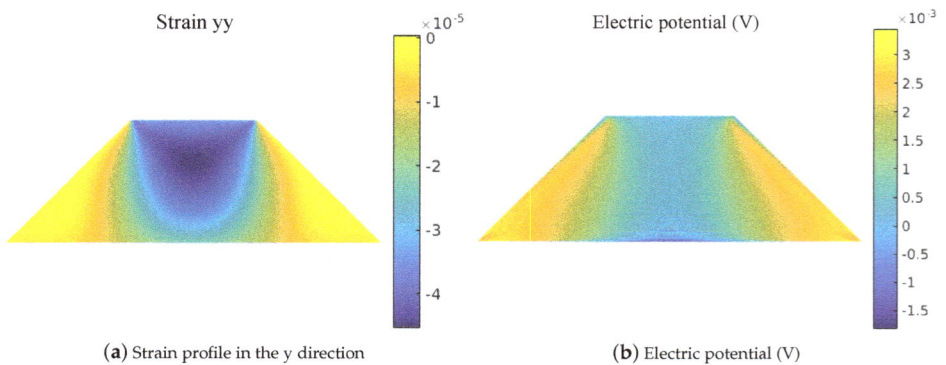

(**a**) Strain profile in the y direction

(**b**) Electric potential (V)

Figure 7. Calculation result of the compressed truncated pyramid: (**a**) strain profile in the y direction; (**b**) electric field profile.

3.3. Flexoelectricity in Composite

In this section, we demonstrate the possibility of inducing electric polarization in a composite system. The composite system is a combination of an embedding matrix in a square shape and a non-piezoelectric material (inclusion) in a circular shape, as shown in Figure 8. The composite under a uniform mechanical loading induces electrical polarization due to the local strain gradients near the material discontinuity. The dielectric constant of the inclusion material is about 10% of the embedding matrix. It is assumed that the Young's modulus of the matrix material (E_{mat}) is lower than the Young's modulus of the inclusion material (E_{inc}). Three different Young's modulus ratios ($\frac{E_{inc}}{E_{mat}} = 10, 100, 1000$) are used to understand its influence on the energy transfer ratio between mechanical energy and electrical energy. The edge length of the square domain is $L = 10$ μm with center inclusion radius $r_1 = 1.5$ μm. The remaining parameters (Table 1) for both materials are similar.

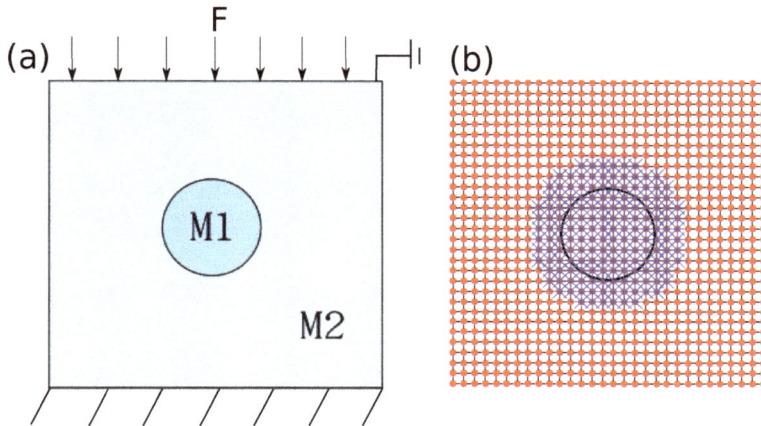

Figure 8. Schematic illustration of: (**a**) the square domain with center inclusion and its boundary condition; (**b**) MLS discretization with the red solid point representing the nodes and the blue asterisk representing the enriched nodes.

Figure 9 presents the strain and electric potential profile for the domain with a center inclusion. The non-uniform strain field near the inclusion boundary generates the polarization gradient and electric potential. Figure 10 plots the strain gradient profile: $\epsilon_{yy,y}$ and $\epsilon_{xx,x}$ along the horizontal and vertical center line of the square domain, respectively. In both directions, a high strain gradient is seen near the boundary of the inclusion, which is due to the different material toughness of the matrix and the inclusion. Since the inclusion material is a non-piezoelectric material, the induced electrical potential is a consequence of flexoelectricity. Therefore, the linear relationship between electrical potential and strain gradient generates the strong potential near the inclusion, as seen from Figure 9b.

In general, the volume percentage of the inclusions has a vast impact on the overall composite properties [56]. In order to investigate the volume percentage effect on the electromechanical coupling, we have repeated the simulations with many inclusions under the same loading condition. The number of inclusions is varied according to the area ratio from 0.5–2.5%. The area ratio is defined as the total area of inclusions divided by the area of a square domain. For each area ratio, several inclusion configurations (center location for inclusions) are examined, and later, the results are averaged. Note that, for inserting many inclusions, the radius is decreased to 0.4 μm. Figure 11a shows the strong electric potential near the inclusion boundary for a domain with 10 randomly-distributed inclusions. This is due to the non-uniform strain distributions near the inclusion boundary. A strong electric potential is noted in the region between the nearby inclusions. This corresponds to the interaction of the non-uniform strain fields around the neighboring inclusions. Figure 11b shows the mesh configuration. Figure 12 plots the variation of the total electrical energy transfer rate with the area ratio. The energy transfer increases with the increase of the inclusion area ratio. This behavior is expected since higher irregularities of strain are induced when many inclusions are present. Figure 12 also indicates that domains with an identical inclusion area ratio with higher $\frac{E_{inc}}{E_{mat}}$ induce higher energy conversion from mechanical to electrical energy. This suggests that softer matrix materials enhance the electromechanical coupling characteristics.

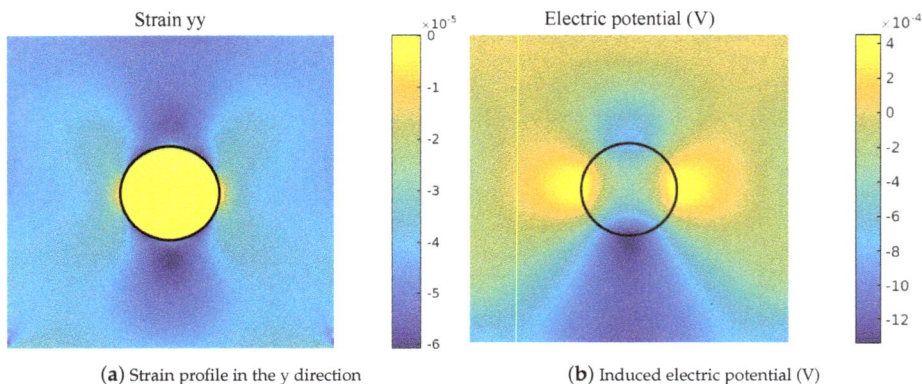

(**a**) Strain profile in the y direction   (**b**) Induced electric potential (V)

Figure 9. Calculation results of a square composition under compression: (**a**) strain profile in the y direction; (**b**) induced electric potential.

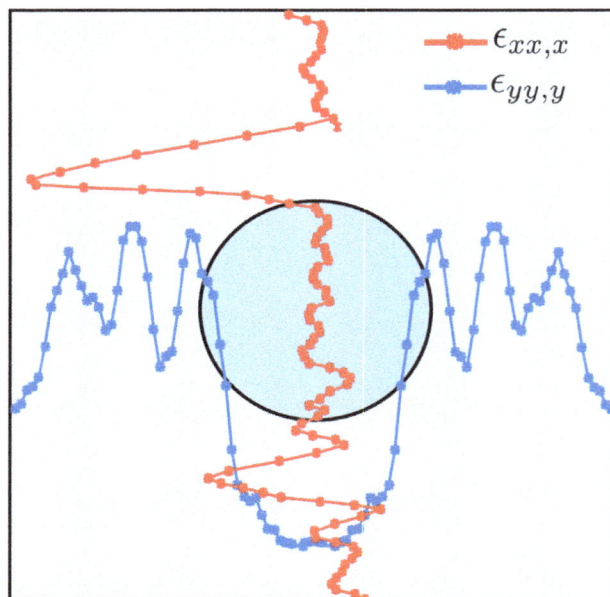

Figure 10. Strain gradient profile at the horizontal center line ($\epsilon_{yy,y}$) and the vertical center line ($\epsilon_{xx,x}$).

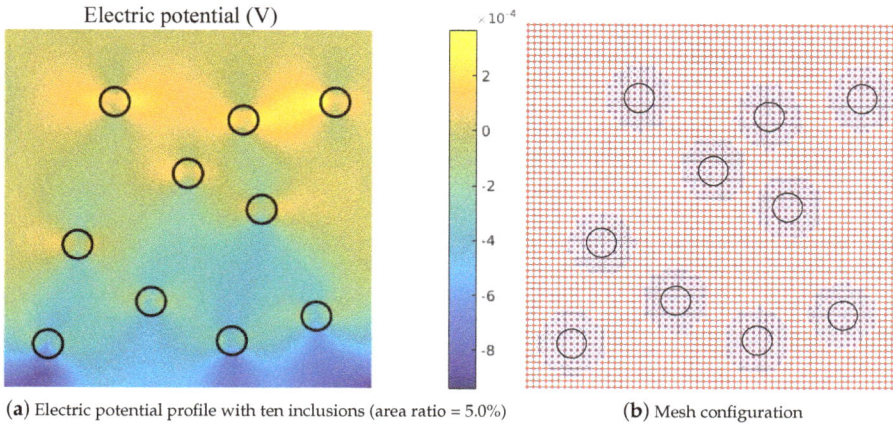

(**a**) Electric potential profile with ten inclusions (area ratio = 5.0%)   (**b**) Mesh configuration

Figure 11. Calculation results of the square domain with randomly-distributed inclusions under compression: (**a**) electric potential profile for the domain with ten inclusions; (**b**) MLS discretization with the red solid point representing the nodes and the blue asterisk representing the enriched nodes.

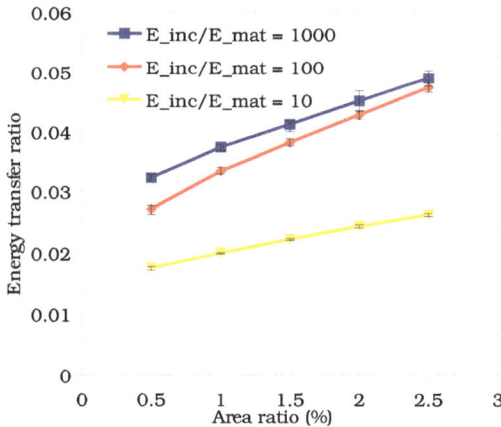

Figure 12. The energy conversion rate calculated by (Equation (14)) vs. the inclusion area ratio. Several setups are created for each configuration; hereafter, the averaged results are plotted. The error bar represents the upper and lower boundary of the calculation results.

4. Conclusions

In this study, the proposed model demonstrates the possibilities of inducing electromechanical coupling in a nano-composite material without the presence of the piezoelectric effect. The electromechanical coupling is modeled using the EFG method. Strain gradient involving partial differential equations is numerically solved using the MLS approximation. The C^1 continuity due to the strain gradient term is fulfilled by choosing the special weight function in the EFG. The level set techniques are employed to handle the weak discontinuity between the inclusion and matrix. The present numerical model captures the finite size flexoelectric effect for a mechanically-loaded cantilever beam. The converse flexoelectric effect-induced mechanical deformation of an electrically loaded cantilever beam is in good agreement with earlier reports. The results of the truncated pyramid further validate the current model. The non-uniform strain fields near the inclusion boundary induce the electrical polarization and thus electric potential due to flexoelectricity. We also found that the magnitude of the electromechanical coupling is largely dependent on the area ratio between the inclusion and the matrix. The higher the inclusion area ratio, the stronger the electromechanical coupling. Furthermore, a softer matrix material can also enhance the electromechanical coupling when compared to a stiff material. These findings help in designing a nano-composite utilizing the flexoelectric effect.

Author Contributions: Conceptualization, B.H.; Methodology, B.H.; Software, B.H.; Validation, B.H.; Formal Analysis, B.H. and B.J.; Investigation, B.H. and B.J.; Resources, B.H. and B.J.; Data Curation, B.H. and B.J.; Writing-Original Draft Preparation, B.H. and B.J.; Writing-Review and Editing, B.H. and B.J.; Visualization, B.H. and B.J.; Supervision, X.Z.; Project Administration, X.Z.; Funding Acquisition, X.Z.

Funding: This research is funded by the National Science Foundation of China (11772234) and ERC Starting Grant 802205.

Conflicts of Interest: Authors declare no conflict of interest.

Appendix A. Theory of Flexoelectricity

The enthalpy density $\mathcal{H}$ for a dielectric solid with piezoelectric and flexoelectric effects is [50,51]:

$$\mathcal{H}(\epsilon_{ij}, E_i, \epsilon_{jk,l}, E_{i,j}) = \frac{1}{2}\mathbb{C}_{ijkl}\epsilon_{ij}\epsilon_{kl} - e_{ikl}E_i\epsilon_{kl} + (d_{ijkl}E_{i,j}\epsilon_{kl} + f_{ijkl}E_i\epsilon_{jk,l}) - \frac{1}{2}\kappa_{ij}E_iE_j, \quad \text{(A1)}$$

where $E_i = -\phi_{,i}$ is the electric field; ϕ_i is the electric potential; ϵ_{ij} is the mechanical strain; $\mathbb{C}_{ijkl}$ is the fourth-order elastic moduli tensor; e_{ikl} is the third-order tensor of piezoelectricity; f_{iljk} and d_{ijkl} are the fourth-order direct and converse flexoelectric tensors; and κ is the second-order dielectric tensor. Sharma et al. [52] defined the flexoelectric tensor μ_{ijkl} as the difference between d_{iljk} and f_{ijkl} by the application of integration by parts and Gauss divergence theorem to Equation (A1), which is:

$$
\begin{aligned}
\int_\Omega \left(d_{ijkl}E_{i,j}\epsilon_{kl} + f_{ijkl}E_i\epsilon_{jk,l} \right) d\Omega &= \int_\Omega d_{ijkl}E_{i,j}\epsilon_{kl}d\Omega + \int_\Omega f_{ijkl}E_i\epsilon_{jk,l}d\Omega \\
&= \int_{\partial\Omega} d_{ijkl}E_i\epsilon_{kl}dS - \int_\Omega d_{ijkl}E_i\epsilon_{kl,j}d\Omega + \int_\Omega f_{ijkl}E_i\epsilon_{jk,l}d\Omega \\
&= -\int_\Omega \left(d_{ijkl}E_i\epsilon_{kl,j} - f_{ijkl}E_i\epsilon_{jk,l} \right) d\Omega + \int_{\partial\Omega} d_{ijkl}E_i\epsilon_{kl}dS \qquad \text{(A2)} \\
&= -\int_\Omega \left(d_{iljk} - f_{ijkl} \right) E_i\epsilon_{jk,l}d\Omega + \int_{\partial\Omega} d_{ijkl}E_i\epsilon_{kl}dS \\
&= -\int_\Omega \mu_{ijkl}E_i\epsilon_{jk,l}d\Omega + \int_{\partial\Omega} d_{ijkl}E_i\epsilon_{kl}dS.
\end{aligned}
$$

Rewriting Equation (A1) leads to Equation (1).

The stress $(\hat{\sigma}_{ij})$ and electric displacement $(\hat{D}_i)$ from Equation (1) when considering the piezoelectricity effect is given as:

$$\hat{\sigma}_{ij} = \frac{\partial\mathcal{H}}{\partial\epsilon_{ij}}; \qquad\qquad \hat{D}_i = -\frac{\partial\mathcal{H}}{\partial E_i}. \qquad \text{(A3)}$$

Due to the presence of flexoelectricity, the higher-order stress $(\bar{\sigma}_{ijk})$ and electric displacement $(\bar{D}_{ij})$ read:

$$\bar{\sigma}_{ijk} = \frac{\partial\mathcal{H}}{\partial\epsilon_{ij,k}}; \qquad\qquad \bar{D}_{ij} = -\frac{\partial\mathcal{H}}{\partial E_{i,j}}. \qquad \text{(A4)}$$

The total stress and electric displacement from piezoelectric and flexoelectric effects are summarized as:

$$
\begin{aligned}
\sigma_{ij} &= \hat{\sigma}_{ij} - \bar{\sigma}_{ijk,k} = \mathbb{C}_{ijkl}\epsilon_{kl} - e_{kij}E_k + \mu_{lijk}E_{l,k}; \\
D_i &= \hat{D}_i - \bar{D}_{ij,j} = e_{ikl}\epsilon_{kl} + \kappa_{ij}E_j + \mu_{ijkl}\epsilon_{jk,l}.
\end{aligned}
\qquad \text{(A5)}
$$

The essential and natural electric boundary condition are:

$$
\begin{aligned}
\phi &= \bar{\phi} && \text{on} && \Gamma_\phi; \\
D_i n_i &= -w && \text{on} && \Gamma_D; \\
\Gamma_\phi \cup \Gamma_D &= \partial\Omega && \text{and} && \Gamma_\phi \cap \Gamma_D = \varnothing,
\end{aligned}
\tag{A6}
$$

where $\bar{\phi}$ and w are the applied electric potential and surface charge density, $\partial\Omega$ represents the boundary of the domain, and n_i is the unit normal to the boundary $\partial\Omega$. The mechanical boundary conditions are given as:

$$
\begin{aligned}
u &= \bar{u} && \text{on} && \Gamma_u; \\
t_k &= \bar{t}_k && \text{on} && \Gamma_t; \\
\Gamma_u \cup \Gamma_t &= \partial\Omega && \text{and} && \Gamma_u \cap \Gamma_t = \varnothing,
\end{aligned}
\tag{A7}
$$

where $\bar{u}$ and $\bar{t}_k$ are prescribed mechanical displacement and traction. The remaining boundary conditions (normal derivation of displacement and higher-order tractions) resulting from the strain gradient have been set to zero under the assumption of a homogeneous condition.

Rewrite Equations (A3) and (A4) as:

$$
\begin{aligned}
\partial\mathcal{H} &= \hat{\sigma}_{ij}\partial\epsilon_{ij} \\
\partial\mathcal{H} &= \bar{\sigma}_{ijk}\partial\epsilon_{ij,k}, \\
\partial\mathcal{H} &= -\hat{D}_i\partial E_i
\end{aligned}
\tag{A8}
$$

and integration over the domain Ω gives:

$$
H = \frac{1}{2}\int_\Omega \left(\hat{\sigma}_{ij}\epsilon_{ij} + \bar{\sigma}_{ijk}\epsilon_{ij,k} - \hat{D}_i E_i\right) d\Omega,
\tag{A9}
$$

where H is the total electrical enthalpy. The external work by surface mechanical and electrical forces is given by:

$$
W_{ext} = \int_{\Gamma_t} \bar{t}_i u_i dS - \int_{\Gamma_D} w\phi dS.
\tag{A10}
$$

Finally, the weak form of the mechanical and electrical equilibrium derived from the Hamilton principle for the static problem yields:

$$
0 = \int_\Omega \left(\hat{\sigma}_{ij}\delta\epsilon_{ij} + \bar{\sigma}_{ijk}\delta\epsilon_{ij,k} - \hat{D}_i\delta E_i\right) d\Omega - \int_{\Gamma_t} \bar{t}_i \delta u_i dS - \int_{\Gamma_D} w\delta\phi dS.
\tag{A11}
$$

Substituting Equations (A3)–(A5) into Equation (A11) yields:

$$
\begin{aligned}
&\int_\Omega \left(\mathbb{C}_{ijkl}\delta\epsilon_{ij}\epsilon_{kl} - e_{kij}E_k\delta\epsilon_{ij} - \mu_{lijk}E_l\delta\epsilon_{ij,k} - \kappa_{ij}\delta E_i E_j - e_{ikl}\delta E_i\epsilon_{kl} - \mu_{ijkl}\delta E_i\epsilon_{jk,l}\right) d\Omega \\
&- \int_{\Gamma_t} \bar{t}_i \delta u_i dS - \int_{\Gamma_D} w\delta\phi dS = 0.
\end{aligned}
\tag{A12}
$$

The unknowns (displacement and electric potential) in Equation (A12) are approximated using MLS.

Appendix B. Details of the Shape Function

The shape function $\Phi_I(x)$ associates with Node I and a point x under MLS is:

$$\Phi_I(x) = \boldsymbol{p}^T(x) \left[\boldsymbol{A}(x)\right]^{-1} w\left(x - x_I\right) \boldsymbol{p}(x_I), \tag{A13}$$

where $\boldsymbol{p}(x)$ is the second-order polynomial, which is:

$$\boldsymbol{p}^T(x) = \begin{bmatrix} 1 & x & y & x^2 & xy & y^2 \end{bmatrix}. \tag{A14}$$

The quadratic spline weight function w ensures C^3 continuity inside an element and C^2 continuity between elements [40]. The mathematical expression for w is:

$$w(r) = \begin{cases} 1 - 6r^2 + 8r^3 - 3r^4 & \text{if } r \leq 1 \\ 0 & \text{if } r > 1 \end{cases} \tag{A15}$$

where:

$$r = \frac{\| x_I - x \|}{d}, \tag{A16}$$

d is the predefined search radius of the support domain and d equals three-times the nodal spacing. The moment matrix $\boldsymbol{A}(x)$ has the form:

$$\boldsymbol{A}(x) = \sum_{I=1}^{N} w(x - x_I)\boldsymbol{p}(x_I)\boldsymbol{p}^T(x_I). \tag{A17}$$

A sufficient number of support nodes ensures the non-singularity of matrix $\boldsymbol{A}(x)$. The enrichment function $\Psi(x)$ has the form [42]:

$$\Psi(x) = \text{abs}(\psi(x)); \qquad \text{where } \psi(x) = \min_{i=1,2,\ldots,n_c} \left\{ \| x - x_c^i \| - r_c^i \right\} \tag{A18}$$

where n_c is the total number of inclusions inside the domain, x_c^i is the center coordinate of the ith circular inclusion, and r_c^i is the radius of the ith circular inclusion.

Appendix C. Mathematical Expression for the Elements in Equation (4)

$$B_u = \partial \Phi_u = \begin{bmatrix} \frac{\partial}{\partial x} & 0 & \frac{\partial}{\partial y} \\ 0 & \frac{\partial}{\partial y} & \frac{\partial}{\partial x} \end{bmatrix}; \tag{A19}$$

$$B_\phi = \partial \Phi_\phi = \begin{bmatrix} \frac{\partial}{\partial x} & \frac{\partial}{\partial y} \end{bmatrix}; \tag{A20}$$

$$H_u = \partial\partial \Phi_u = \begin{bmatrix} \frac{\partial^2}{\partial x^2} & 0 & \frac{\partial^2}{\partial x\partial y} & \frac{\partial^2}{\partial x\partial y} & 0 & \frac{\partial^2}{\partial y^2} \\ 0 & \frac{\partial^2}{\partial x\partial y} & \frac{\partial^2}{\partial x^2} & 0 & \frac{\partial^2}{\partial y^2} & \frac{\partial^2}{\partial x\partial y} \end{bmatrix}; \tag{A21}$$

$$B_u^{enr} = \partial\Phi_u\Psi + \Phi_u\partial\Psi; \tag{A22}$$

$$B_\phi^{enr} = \partial\Phi_\phi\Psi + \Phi_\phi\partial\Psi; \tag{A23}$$

$$H_u^{enr} = \partial\partial\Phi_u\Psi + \Phi_u\partial\partial\Psi + 2\partial\Phi_u\partial\Psi; \tag{A24}$$

$$
\mathbf{B}_u^{enr} = \begin{bmatrix} \frac{\partial}{\partial x}\psi(x) + \mathrm{sign}(\psi(x))\frac{(x-x_c)}{d} & 0 \\[4pt] 0 & \frac{\partial}{\partial x}\psi(x) + \mathrm{sign}(\psi(x))\frac{(y-y_c)}{d} \\[4pt] \frac{\partial}{\partial x}\psi(x) + \mathrm{sign}(\psi(x))\frac{(y-y_c)}{d} & \frac{\partial}{\partial x}\psi(x) + \mathrm{sign}(\psi(x))\frac{(x-x_c)}{d} \end{bmatrix}^T ;
$$

(A25)

$$
\mathbf{B}_\phi^{enr} = \begin{bmatrix} \frac{\partial}{\partial x}\psi(x) + \mathrm{sign}(\psi(x))\frac{(x-x_c)}{d} \\[6pt] \frac{\partial}{\partial x}\psi(x) + \mathrm{sign}(\psi(x))\frac{(y-y_c)}{d} \end{bmatrix}^T ;
$$

(A26)

$$
\mathbf{H}_u^{enr} = \left[\begin{array}{cc}
\frac{\partial^2}{\partial y \partial x}\psi(x) + \frac{\partial}{\partial y}\mathrm{sign}(\psi(x))\frac{(x-x_c)}{d} + \frac{\partial}{\partial x}\mathrm{sign}(\psi(x))\frac{(y-y_c)}{d} + \mathrm{sign}(\psi(x))\frac{-(x-x_c)*(y-y_c)}{d^{(\frac{3}{2})}} & 0 \\
\frac{\partial^2}{\partial y \partial x}\psi(x) + \frac{\partial}{\partial x}\mathrm{sign}(\psi(x))\frac{(x-x_c)}{d} + \frac{\partial}{\partial x}\mathrm{sign}(\psi(x))\frac{(y-y_c)}{d} + \mathrm{sign}(\psi(x))\frac{-(x-x_c)*(y-y_c)}{d^{(\frac{3}{2})}} & 0 \\
\frac{\partial^2}{\partial x^2}\psi(x) + 2\frac{\partial}{\partial x}\mathrm{sign}(\psi(x))\frac{(x-x_c)}{d} + \mathrm{sign}(\psi(x))\frac{(y-y_c)^2}{d^{(\frac{3}{2})}} & 0 \\
\frac{\partial^2}{\partial y^2}\psi(x) + 2\frac{\partial}{\partial y}\mathrm{sign}(\psi(x))\frac{(x-x_c)}{d} + \mathrm{sign}(\psi(x))\frac{(y-y_c)}{d} & 0 \\
0 & \frac{\partial^2}{\partial y \partial x}\psi(x) + \frac{\partial}{\partial y}\mathrm{sign}(\psi(x))\frac{(x-x_c)}{d} + \frac{\partial}{\partial x}\mathrm{sign}(\psi(x))\frac{(y-y_c)}{d} + \mathrm{sign}(\psi(x))\frac{-(x-x_c)*(y-y_c)}{d^{(\frac{3}{2})}} \\
0 & \frac{\partial^2}{\partial x^2}\psi(x) + \frac{\partial}{\partial y}\mathrm{sign}(\psi(x))\frac{(x-x_c)}{d} + \frac{\partial}{\partial x}\mathrm{sign}(\psi(x))\frac{(y-y_c)}{d} + \mathrm{sign}(\psi(x))\frac{(x-x_c)^2}{d^{(\frac{3}{2})}} \\
0 & \frac{\partial^2}{\partial y \partial x}\psi(x) + \frac{\partial}{\partial y}\mathrm{sign}(\psi(x))\frac{(y-y_c)}{d} + \mathrm{sign}(\psi(x))\frac{-(x-x_c)*(y-y_c)}{d^{(\frac{3}{2})}}
\end{array} \right]^T .
$$

(A27)

References

1. Chen, J.; Qiu, Q.; Han, Y.; Lau, D. Piezoelectric materials for sustainable building structures: Fundamentals and applications. *Renew. Sustain. Energy Rev.* **2019**, *101*, 14–25. [CrossRef]
2. Cabeza, L.F.; Barreneche, C.; Miró, L.; Morera, J.M.; Bartolí, E.; Fernández, A.I. Low carbon and low embodied energy materials in buildings: A review. *Renew. Sustain. Energy Rev.* **2013**, *23*, 536–542. [CrossRef]
3. Ramesh, T.; Prakash, R.; Shukla, K. Life cycle energy analysis of buildings: An overview. *Energy Build.* **2010**, *42*, 1592–1600. [CrossRef]
4. Scopelianos, A.G. Piezoelectric Biomedical Device. US Patent 5,522,879, 4 June 1996.
5. Zhou, Q.; Lau, S.; Wu, D.; Shung, K.K. Piezoelectric films for high frequency ultrasonic transducers in biomedical applications. *Prog. Mater. Sci.* **2011**, *56*, 139–174. [CrossRef] [PubMed]
6. Jiang, X.; Huang, W.; Zhang, S. Flexoelectric nano-generator: Materials, structures and devices. *Nano Energy* **2013**, *2*, 1079–1092. [CrossRef]
7. Zhuang, X.Y.; Huang, R.Q.; Liang, C.; Rabczuk, T. A coupled thermo-hydro-mechanical model of jointed hard rock for compressed air energy storage. *Math. Probl. Eng.* **2014**, *2014*, 179169. [CrossRef]
8. Ahmadpoor, F.; Sharma, P. Flexoelectricity in two-dimensional crystalline and biological membranes. *Nanoscale* **2015**, *7*, 16555–16570. [CrossRef]
9. Ma, W.; Cross, L.E. Observation of the flexoelectric effect in relaxor Pb (Mg 1/3 Nb 2/3) O 3 ceramics. *Appl. Phys. Lett.* **2001**, *78*, 2920–2921. [CrossRef]
10. Ma, W.; Cross, L.E. Flexoelectricity of barium titanate. *Appl. Phys. Lett.* **2006**, *88*, 232902. [CrossRef]
11. Hong, J.; Catalan, G.; Scott, J.; Artacho, E. The flexoelectricity of barium and strontium titanates from first principles. *J. Phys. Condens. Matter* **2010**, *22*, 112201. [CrossRef]
12. Xu, T.; Wang, J.; Shimada, T.; Kitamura, T. Direct approach for flexoelectricity from first-principles calculations: Cases for SrTiO3 and BaTiO3. *J. Phys. Condens. Matter* **2013**, *25*, 415901. [CrossRef] [PubMed]
13. Kundalwal, S.; Meguid, S.; Weng, G. Strain gradient polarization in graphene. *Carbon* **2017**, *117*, 462–472. [CrossRef]
14. Javvaji, B.; He, B.; Zhuang, X. The generation of piezoelectricity and flexoelectricity in graphene by breaking the materials symmetries. *Nanotechnology* **2018**, *29*, 225702. [CrossRef] [PubMed]
15. He, B.; Javvaji, B.; Zhuang, X. Size dependent flexoelectric and mechanical properties of barium titanate nanobelt: A molecular dynamics study. *Phys. B Condens. Matter* **2018**, *545*, 527–535. [CrossRef]
16. Zhang, Z.; Geng, D.; Wang, X. Calculation of the piezoelectric and flexoelectric effects in nanowires using a decoupled finite element analysis method. *J. Appl. Phys.* **2016**, *119*, 154104. [CrossRef]
17. Nanthakumar, S.; Zhuang, X.; Park, H.S.; Rabczuk, T. Topology optimization of flexoelectric structures. *J. Mech. Phys. Solids* **2017**, *105*, 217–234. [CrossRef]
18. Nanthakumar, S.S.; Lahmer, T.; Zhuang, X.Y.; Zi, G.; Rabczuk, T. Detection of material interfaces using a regularized level set method in piezoelectric structures. *Inverse Probl. Sci. Eng.* **2016**, *24*, 153–176. [CrossRef]
19. Deng, F.; Deng, Q.; Yu, W.; Shen, S. Mixed Finite Elements for Flexoelectric Solids. *J. Appl. Mech.* **2017**, *84*, 081004. [CrossRef]
20. Mao, S.; Purohit, P.K.; Aravas, N. Mixed finite-element formulations in piezoelectricity and flexoelectricity. *Proc. R. Soc. A* **2016**, *472*, 20150879. [CrossRef]
21. Ghasemi, H.; Park, H.S.; Rabczuk, T. A level-set based IGA formulation for topology optimization of flexoelectric materials. *Comput. Methods Appl. Mech. Eng.* **2017**, *313*, 239–258. [CrossRef]
22. Hamdia, K.M.; Ghasemi, H.; Zhuang, X.Y.; Alajlan, N.; Rabczuk, T. Sensitivity and uncertainty analysis for flexoelectric nanostructures. *Comput. Methods Appl. Mech. Eng.* **2018**, *337*, 95–109. [CrossRef]
23. Nguyen-Thanh, N.; Valizadeh, N.; Nguyen, M.N.; Nguyen-Xuan, H.; Zhuang, X.Y.; Areias, P.; Zi, G.; Bazilevs, Y.; De Lorenzis, L.; Rabczuk, T. An extended isogeometric thin shell analysis based on Kirchhoff–Love theory. *Comput. Methods Appl. Mech. Eng.* **2015**, *284*, 265–291. [CrossRef]
24. Nguyen, V.P.; Anitescu, C.; Bordas, S.P.; Rabczuk, T. Isogeometric analysis: An overview and computer implementation aspects. *Math. Comput. Simul.* **2015**, *117*, 89–116. [CrossRef]
25. Hughes, T.J.; Cottrell, J.A.; Bazilevs, Y. Isogeometric analysis: CAD, finite elements, NURBS, exact geometry and mesh refinement. *Comput. Methods Appl. Mech. Eng.* **2005**, *194*, 4135–4195. [CrossRef]
26. Abdollahi, A.; Peco, C.; Millán, D.; Arroyo, M.; Arias, I. Computational evaluation of the flexoelectric effect in dielectric solids. *J. Appl. Phys.* **2014**, *116*, 093502. [CrossRef]

27. Abdollahi, A.; Millán, D.; Peco, C.; Arroyo, M.; Arias, I. Revisiting pyramid compression to quantify flexoelectricity: A three-dimensional simulation study. *Phys. Rev. B* **2015**, *91*, 104103. [CrossRef]

28. Bernholc, J.; Nakhmanson, S.M.; Nardelli, M.B.; Meunier, V. Understanding and enhancing polarization in complex materials. *Comput. Sci. Eng.* **2004**, *6*, 12–21. [CrossRef]

29. Ghidelli, M.; Sebastiani, M.; Collet, C.; Guillemet, R. Determination of the elastic moduli and residual stresses of freestanding Au-TiW bilayer thin films by nanoindentation. *Mater. Des.* **2016**, *106*, 436–445. [CrossRef]

30. Msekh, M.A.; Cuong, N.H.; Zi, G.; Areias, P.; Zhuang, X.Y.; Rabczuk, T. Fracture properties prediction of clay/epoxy nanocomposites with interphase zones using a phase field model. *Eng. Fract. Mech.* **2018**, *188*, 287–299. [CrossRef]

31. Bhavanasi, V.; Kumar, V.; Parida, K.; Wang, J.; Lee, P.S. Enhanced Piezoelectric Energy Harvesting Performance of Flexible PVDF-TrFE Bilayer Films with Graphene Oxide. *ACS Appl. Mater. Interfaces* **2016**, *8*, 521–529. [CrossRef]

32. Biswal, A.K.; Das, S.; Roy, A. Designing and synthesis of a polymer matrix piezoelectric composite for energy harvesting. *IOP Conf. Ser. Mater. Sci. Eng.* **2017**, *178*, 012002, doi:10.1088/1742-6596/755/1/011001. [CrossRef]

33. Sidhardh, S.; Ray, M.C. Effective properties of flexoelectric fiber-reinforced nanocomposite. *Mater. Today Commun.* **2018**, *17*, 114–123. [CrossRef]

34. Hamdia, K.M.; Silani, M.; Zhuang, X.Y.; He, P.F.; Rabczuk, T. Stochastic analysis of the fracture toughness of polymeric nanoparticle composites using polynomial chaos expansions. *Int. J. Fract.* **2017**, *206*, 215–227. [CrossRef]

35. Ghasemi, H.; Park, H.S.; Rabczuk, T. A multi-material level set-based topology optimization of flexoelectric composites. *Comput. Methods Appl. Mech. Eng.* **2018**, *332*, 47–62. [CrossRef]

36. Cordes, L.; Moran, B. Treatment of material discontinuity in the element-free Galerkin method. *Comput. Methods Appl. Mech. Eng.* **1996**, *139*, 75–89. [CrossRef]

37. Rabczuk, T.; Belytschko, T.; Xiao, S.P. Stable particle methods based on Lagrangian kernels. *Comput. Methods Appl. Mech. Eng.* **2004**, *193*, 1035–1063. [CrossRef]

38. Krongauz, Y.; Belytschko, T. EFG approximation with discontinuous derivatives. *Int. J. Numer. Methods Eng.* **1998**, *41*, 1215–1233. [CrossRef]

39. Rabczuk, T.; Zi, G.; Bordas, S.; Nguyen-Xuan, H. A geometrically non-linear three-dimensional cohesive crack method for reinforced concrete structures. *Eng. Fract. Mech.* **2008**, *75*, 4740–4758. [CrossRef]

40. Nguyen, V.P.; Rabczuk, T.; Bordas, S.; Duflot, M. Meshless methods: A review and computer implementation aspects. *Math. Comput. Simul.* **2008**, *79*, 763–813. [CrossRef]

41. Budarapu, P.R.; Gracie, R.; Yang, S.W.; Zhuang, X.Y.; Rabczuk, T. Efficient coarse graining in multiscale modeling of fracture. *Theor. Appl. Fract. Mech.* **2014**, *69*, 126–143. [CrossRef]

42. Sukumar, N.; Chopp, D.L.; Moës, N.; Belytschko, T. Modeling holes and inclusions by level sets in the extended finite-element method. *Comput. Methods Appl. Mech. Eng.* **2001**, *190*, 6183–6200. [CrossRef]

43. Ren, H.L.; Zhuang, X.Y.; Rabczuk, T. Dual-horizon peridynamics: A stable solution to varying horizons. *Comput. Methods Appl. Mech. Eng.* **2017**, *318*, 762–782. [CrossRef]

44. Ren, H.L.; Zhuang, X.Y.; Cai, Y.C.; Rabczuk, T. Dual-horizon peridynamics. *Int. J. Numer. Methods Eng.* **2016**, *108*, 1451–1476. [CrossRef]

45. Rabczuk, T.; Bordas, S.; Zi, G. On three-dimensional modelling of crack growth using partition of unity methods. *Comput. Struct.* **2010**, *88*, 1391–1411. [CrossRef]

46. Rabczuk, T.; Areias, P.M.A.; Belytschko, T. A meshfree thin shell method for non-linear dynamic fracture. *Int. J. Numer. Methods Eng.* **2007**, *72*, 524–548. [CrossRef]

47. Rabczuk, T.; Belytschko, T. A three-dimensional large deformation meshfree method for arbitrary evolving cracks. *Comput. Methods Appl. Mech. Eng.* **2007**, *196*, 2777–2799. [CrossRef]

48. Rabczuk, T.; Belytschko, T. Cracking particles: A simplified meshfree method for arbitrary evolving cracks. *Int. J. Numer. Methods Eng.* **2004**, *61*, 2316–2343. [CrossRef]

49. Rabczuk, T.; Bordas, S.; Zi, G. A three-dimensional meshfree method for continuous multiple-crack initiation, propagation and junction in statics and dynamics. *Comput. Mech.* **2007**, *40*, 473–495. [CrossRef]

50. Majdoub, M.; Sharma, P.; Cagin, T. Enhanced size-dependent piezoelectricity and elasticity in nanostructures due to the flexoelectric effect. *Phys. Rev. B* **2008**, *77*, 125424. [CrossRef]

51. Majdoub, M.S.; Sharma, P.; Çağin, T. Erratum: Enhanced size-dependent piezoelectricity and elasticity in nanostructures due to the flexoelectric effect [Phys. Rev. B 77, 125424 (2008)]. *Phys. Rev. B* **2009**, *79*, 119904, doi:10.1103/PhysRevB.79.119904. [CrossRef]

52. Sharma, N.; Landis, C.; Sharma, P. Piezoelectric thin-film superlattices without using piezoelectric materials. *J. Appl. Phys.* **2010**, *108*, 024304. [CrossRef]

53. Belytschko, T.; Lu, Y.Y.; Gu, L. Element-free Galerkin methods. *Int. J. Numer. Methods Eng.* **1994**, *37*, 229–256. [CrossRef]

54. Belytschko, T.; Lu, Y.; Gu, L.; Tabbara, M. Element-free Galerkin methods for static and dynamic fracture. *Int. J. Solids Struct.* **1995**, *32*, 2547–2570. [CrossRef]

55. Belytschko, T.; Black, T. Elastic crack growth in finite elements with minimal remeshing. *Int. J. Numer. Methods Eng.* **1999**, *45*, 601–620. [CrossRef]

56. He, B.; Mortazavi, B.; Zhuang, X.; Rabczuk, T. Modeling Kapitza resistance of two-phase composite material. *Compos. Struct.* **2016**, *152*, 939–946. [CrossRef]

energies

MDPI

Article

Crack Patterns in Heterogenous Rocks Using a Combined Phase Field-Cohesive Interface Modeling Approach: A Numerical Study

José Reinoso [1,*], Percy Durand [2], Pattabhi Ramaiah Budarapu [3] and Marco Paggi [4]

[1] Elasticity and Strength of Materials Group, School of Engineering, Universidad de Sevilla, 41092 Seville, Spain
[2] Department of Building Structures and Geotechnical Engineering, Universidad de Sevilla, 41012 Seville, Spain; percy@us.es
[3] School of Mechanical Sciences, Indian Institute of Technology, Bhubaneswar 752050, India; pattabhi@iitbbs.ac.in
[4] IMT School for Advanced Studies Lucca, Piazza San Francesco 19, 55100 Lucca, Italy; marco.paggi@imtlucca.it
* Correspondence: jreinoso@us.es

Received: 26 November 2018; Accepted: 4 March 2019; Published: 13 March 2019

Abstract: Rock fracture in geo-materials is a complex phenomenon due to its intrinsic characteristics and the potential external loading conditions. As a result, these materials can experience intricate fracture patterns endowing various cracking phenomena such as: branching, coalescence, shielding, and amplification, among many others. In this article, we present a numerical investigation concerning the applicability of an original bulk-interface fracture simulation technique to trigger such phenomena within the context of the phase field approach for fracture. In particular, the prediction of failure patterns in heterogenous rock masses with brittle response is accomplished through the current methodology by combining the phase field approach for intact rock failure and the cohesive interface-like modeling approach for its application in joint fracture. Predictions from the present technique are first validated against Brazilian test results, which were developed using alternative phase field methods, and with respect to specimens subjected to different loading case and whose corresponding definitions are characterized by the presence of single and multiple flaws. Subsequently, the numerical study is extended to the analysis of heterogeneous rock masses including joints that separate different potential lithologies, leading to tortuous crack paths, which are observed in many practical situations.

Keywords: rock mechanics; phase field approach to fracture; fracture of geo-materials; cohesive zone model; interface modeling

1. Introduction

Fracture events in geological materials are phenomena of notable importance for the safety and stability of geological masses. This is a relevant issue to engineering applications including tunneling procedures, hydraulic fracture for energy reservoirs, deep foundations, to name a few. Due to the advent of novel modeling techniques and the advances in the computational capabilities, the usage of numerical methods for reliable predictions of failure in rock masses has suffered a tremendous improvement in the last few years, becoming a plausible alternative for the analysis of geo-materials.

In this setting, the presence of flaws such as defects/discontinuities in rocks leads to a substantial reduction of the strength of rock mass, as compared to the intact rock. In line with linear elastic fracture mechanics (LEFM), the singular character of the stress field around flaws can induce the propagation of such defects when the stress state within the rock mass is altered, for instance due to a tunnel excavation. Therefore, the existence, initiation, and propagation of fracture not only affect the mechanical response of the rock mass, but also the construction of the engineering structures. This dependence of the strength and deformability of the rock mass upon the intact rock and the characteristic discontinuities (fracture and joints) make the deep understanding of such fracture processes a challenging task.

Stemming from the complex nature of the rock mass, according to [1,2], it is possible to distinguish between crack initiation and propagation phenomena. Crack initiation is defined as the process through which pre-existing flaws initiate in a confined region in the neighborhood of a crack tip. Crack propagation events are those processes by which the pre-existing cracks continue to grow after the initiation. Thus, propagation phases can lead to crack coalescence and/or crack branching or bifurcation, among different scenarios, depending on the characteristics of the rock mass and the external loading conditions. With the aim of achieving a deep understanding of these phenomena, a great deal of research has been conducted in literature to characterize the crack propagation in rocks, which are succinctly discussed from the experimental and numerical standpoints in Sections 1.1 and 1.2, respectively.

1.1. Experimental Investigations

In a landmark investigation, Lajtai [3] performed numerous experimental tests on gypsum of Paris specimens, with single flaws at different inclinations with respect to the direction of the external load. These tests led to the identification of different fracture stages of the rock mass under uniaxial compression. Subsequently Ingraffea et al., [4] have extended the experiments on Limestone and Granodiorite specimens, by orienting the flaws at a particular angle. In [4], the authors reported that the completed crack growth sequence upon failure, which allowed the identification of a complete variety of failure modes. Some recent studies on rocks with several flaws, which were oriented following different inclinations, manifested the complex nature of fracture events in rocks with evident crack coalescence [5–8].

Recently, several researchers have studied the response of rock mass on crack arrest, considering: (*i*) the number of defects, (*ii*) the relative inclination of the flaws and (*iii*) the applied external loads. Mentionable studies on experimental methods have thoroughly investigated uniaxial and biaxial compression tests in rocks, through which it is possible to identify that these materials may experience a combination of flaw slippage onset and wing propagation, followed by secondary cracks (shear dominated) leading to subsequent potential coalescence [5–10]. In particular, Sagong and Bobet [11] distinguished 9 coalescence types of cracks in rock masses.

Nevertheless, it is worth mentioning that the experimental studies on this topic have been profusely exploited in recent decades. Standard methods are often limited due to notable difficulties in the simulation of general loading tests and the preparation of specimens for this purpose. Under these circumstances, traditional experimental procedures are also complemented with the use of digital image correlation (DIC) techniques and 3D printing methods [12], fostering a new perspective for the generation of specimens with desired attributes.

1.2. Numerical Investigations

As mentioned above, many of the previous experimental studies were focused on uniaxial or biaxial compressive loads, since the general loading conditions are difficult to reproduce in the laboratory. This fact motivated the development of computational methods which can potentially capture complex failure mechanisms in rocks. Moreover, numerical methods generally lead to notable cost savings, as compared to arduous experimental campaigns.

Modeling the physical behavior of fracture phenomena in geo-materials has been traditionally tackled using linear elastic fracture mechanics-based methodologies, due to their relative simplicity and representativeness, provided that the confining pressure and loading rate are sufficiently low. From the modeling point of view, previous experimental studies have been employed as reference investigations to calibrate the numerical methods and to provide a deeper insight into the physical processes associated with fracture events in rocks. However, several difficulties have been identified in order to capture the wide variety of failure modes of such materials under different loading cases such as bidirectional traction-compression, traction-shear loading among many others, which evolve from the complex crack path when it is unknown a priori. In addition, tracking such crack paths will be further complicated by the development of branching and coalescence evolutions, which are typical scenarios encountered in geo-materials.

For this purpose, numerous computational methods have been proposed in the last three decades, especially within the context of nonlinear finite element method (FEM) and boundary element method (BEM), to simulate the complex failure modes in rocks. Later on, advanced computational approaches such as: cracking particles method (CPM) [13–15], Peridynamics [16] and dual-horizon Peridynamics [17,18] are developed, which neither do not require an explicit representation of the discontinuities in the displacement field due to fracture nor crack tracking algorithms for the identification of the propagation path.

Exploiting the BEM-based techniques, Einstein and coauthors [19,20] have developed reliable modeling tools to predict crack growth under the loading conditions leading to fracture in Modes I and II, considering the specimens with one or more internal flaws. However, due to the inherent versatility of FEM-based techniques, they are often preferred for fracture applications over alternative methodologies. Within the current modeling techniques to simulate fracture events in solids using FEM, the following categories can be differentiated:

- Methods based on continuum damage mechanics (CDM) encompass the stiffness degradation, by defining a set of internal-damage variables [21]. In their local forms, CDM models suffer from well-known pathological sensitivity of the corresponding estimations on the spatial discretization (finite element mesh) due to the loss of ellipticity of the corresponding governing equations. To mitigate such deficiency, integral regularization schemes and gradient-enhanced formulations have been proposed [22–26], apart from the alternative procedures introduced by Areias and coauthors [27] by combining the local damage formulation with local re-meshing techniques.
- Explicit crack tools can be employed to envisage a strong discontinuity in the displacement field. They rely on either the enrichment of the nodal displacement field at the element level using the partition of unity methods (PUM) [28–30], or the enrichment of the strain field (enhanced FEM, E-FEM) [31–34].
- Adaptive insertion of interface-like cohesive elements within the delimiting edges of the existing mesh, using extrinsic or intrinsic representations of the cohesive law [35–41].

Although the above-mentioned methods have been extensively employed to simulate fracture events in different engineering applications, they present several limitations, particularly to capture the onset of multiple cracks and their evolution, apart from tracking the intricate fracture scenarios including branching and coalescence. This latter aspect is of remarkable complexity in strong discontinuity techniques, which suffer from topological operative problems under such conditions.

Due to the advances in the computational capabilities in the last decade, a renovated interest on diffusive damage models complying with nonlocal representation has been witnessed, which endow a scalar-based evolution equation for the damage field. Especially suitable for brittle fracture events, the so-called phase field (PF) approach of fracture envisages a diffusive crack representation with a nonlocal character [42]. This new concept shares some common aspects with respect to CDM formulations via the definition of a scalar PF variable, which varies from 0 to 1, and triggers the stiffness degradation upon failure.

PF methods can be understood as a regularized version of the Griffith fracture approach [43], through the advocation of the Γ-convergence [44–46], whereby a direct competition between the elastic and the fracture energies is constructed in an integrated global functional of the system. The corresponding minimization of such functional allows capturing the initiation, growth, and propagation of cracks in solids in a robust and reliable manner. Notable contributions in this direction are due to Francfort and Marigo and their coauthors from a fundamental standpoint, see [42,47] and the associated references. Additional formulations exploiting the PF approach for brittle fracture have been rigorously developed by Bourdin et al. [47,48]. Whereas, several modified models have been recently proposed complying with a thermodynamically consistent framework, e.g., see [49–52]. Exploiting this potential, posterior studies, the PF approach has been extended to fracture studies in thin-walled structures [53–55], ductile failure [56,57], cohesive fracture [58], dynamic applications [59,60], multi-physics environments [61–63]. Recently, this methodology has also been applied to study the initiation of fracture in rocks with multiple flaws [64–66].

1.3. Objectives and Organization

Due to the appealing features of the PF method for fracture, the main objective of this work is focused on the application of the PF method and its combination with the interface-like cohesive method to model fracture events in heterogeneous rocks [67]. The advocation of the methodology proposed in [68], which was originally developed by the authors, allows the possible interaction of rock masses with different lithologies, which are separated by a series of predefined discontinuities. The analysis and understanding of interfaces in rocks have always been the subject of a great impact in the research community due to their importance in practice. This stems from the fact that they possess a certain amount of cohesion due to interlocking and roughness, along with frictional responses in case of large sliding modifying the response of the rock mass [69]. Therefore, the aim of this research is to address the potential predictive capabilities of the combined PF-CZM for its application in rock fracture. Based on this interest, the proposed technique is first validated by reproducing experimental results concerning fracture of intact rock and subsequently extended to the analysis of heterogeneous media.

The manuscript is organized as follows: Section 2 summarizes the basic concepts of the PF method for bulk fracture, whereas, the corresponding simulations are presented in Section 3. Posteriorly, general details concerning the combined PF-CZM for triggering fracture events in heterogenous rock mass is outlined in Section 4. Some representative applications to illustrate the applicability of the proposed method are presented in Section 5. Finally, the main conclusions of the present study are given in Section 6.

2. Fundamentals of the Phase Field Approach for Brittle Fracture

The basic concepts of the PF approach for fracture in rocks are briefly introduced in this section. The current formulation advocates the methodology proposed in [47,49], therefore, detailed derivations are herewith omitted for brevity.

2.1. Basic Concepts

Within the context of multi-dimensional analysis, see Figure 1, we consider an arbitrary body $\Omega \in \mathbb{R}^{nd}$ in the Euclidean space of dimension n_d, whose material points are denoted by the position vectors **x** in the Cartesian setting. Body actions are identified by the vector $f_v : \Omega \longrightarrow \mathbb{R}^{n_d}$. The delimiting boundary Ω is defined such that $\partial\Omega \in \mathbb{R}^{n_d-1}$, where $\partial\Omega_u$ and $\partial\Omega_t$ denote the boundary portions in which kinematic and static surface actions are prescribed, respectively, complying to: $\partial\Omega_t \cup \partial\Omega_u = \partial\Omega$ and $\partial\Omega_t \cap \partial\Omega_u = \emptyset$. The kinematic field is denoted by **u**, whereas the second-order Cauchy stress tensor is identified by σ. Boundary conditions in this arbitrary body render:

$$\mathbf{u} = \bar{\mathbf{u}} \quad \text{on } \partial\Omega_u, \quad \text{and} \quad \bar{\mathbf{t}} = \sigma \cdot \mathbf{n} \quad \text{on } \partial\Omega_t, \tag{1}$$

where **n** is the outward normal to the body.

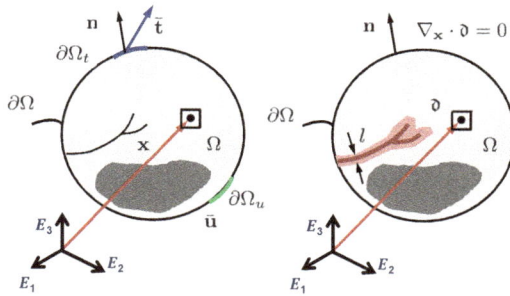

Figure 1. Regularization of the sharp crack topology within the spirit of the phase field approach for fracture.

The PF approach for fracture envisages the definition of a scalar-based phase field variable $\eth$, where $\eth : \Omega \times [0, t] \longrightarrow [0, 1]$ [48], which quantifies the stiffness degradation at the material point level, in order to smear out the sharp crack topology. The intact and fully deteriorated states are identified based on the value of the PF parameter, i.e., $\eth = 0$ and $\eth = 1$, respectively. Therefore, $\eth$ adopts continuous values between 0 and 1. The one-dimensional PF approximation is given by the expression: $\eth = e^{-|x|/l}$, where l indicates the characteristic material length related to the material strength and governs the regularization width. According to [49], the PF problem is governed by:

$$\eth - l^2 \Delta \eth = 0 \quad \text{in } \Omega \quad \text{and} \quad \nabla_x \eth \cdot \mathbf{n} = 0 \quad \text{in } \partial\Omega \tag{2}$$

where $\Delta \eth$ is the Laplacian of $\eth$, and $\nabla_x \eth$ stands for the spatial gradient. At this point, relevant contributions on this matter are due to the contributions by Wu and coauthors [70–72], who extended the applicability the PF method for cohesive and length-scale insensitive fracture.

To account for the regularization, a crack density functional $\gamma(\eth, \nabla_x \eth)$ is defined, which can be estimated using the relation:

$$\Gamma_c(\eth) := \int_\Omega \gamma(\eth, \nabla_x \eth) \, d\Omega. \tag{3}$$

As discussed in [42,45,73], Equation (3) represents the crack surface Γ_c, in the limit $l \to 0$. Furthermore, to prevent crack healing upon unloading, the thermodynamic consistency condition given by $\dot{\Gamma}_c(\eth) \geq 0$, is required. The crack density functional can be then expressed as [49]:

$$\gamma(\eth, \nabla_x \eth) = \frac{1}{2l} \eth^2 + \frac{l}{2} |\nabla_x \eth|^2 . \tag{4}$$

The above definition of $\gamma(\eth, \nabla_x \eth)$ allows the approximation of the crack surface integral in the original Griffith formulation by a volume integral as follows:

$$\int_{\Gamma_c} \mathcal{G}_c \, d\partial\Omega \approx \int_\Omega \mathcal{G}_c \gamma(\eth, \nabla_x \eth) \, d\Omega. \tag{5}$$

Therefore, the total potential of the body can be split into the internal and external counterparts represented, $\Pi_{int}(\mathbf{u}, \eth)$ and $\Pi_{ext}(\mathbf{u})$, respectively:

$$\Pi(\mathbf{u}, \eth) = \Pi_{int}(\mathbf{u}, \eth) + \Pi_{ext}(\mathbf{u}) = \int_\Omega \psi(\varepsilon, \eth) \, d\Omega + \int_\Omega \mathcal{G}_c \gamma(\eth, \nabla_x \eth) \, d\Omega + \Pi_{ext}(\mathbf{u}), \tag{6}$$

where the first term in the integral represents the elastic energy of the body and the second term accounts for the energy dissipation.

To simulate the crack growth under tensile load, we adopt the spectral decomposition of the infinitesimal strain tensor between its corresponding positive and negative counterparts [49]:

$$\psi(\varepsilon, \eth) = \mathfrak{g}(\eth)\psi^e_+(\varepsilon) + \psi^e_-(\varepsilon), \tag{7a}$$

$$\psi^e_+(\varepsilon) = \frac{\lambda}{2}\left(\langle \mathrm{tr}[\varepsilon]\rangle_+\right)^2 + \mu\mathrm{tr}[\varepsilon^2_+], \tag{7b}$$

$$\psi^e_-(\varepsilon) = \frac{\lambda}{2}\left(\langle \mathrm{tr}[\varepsilon]\rangle_-\right)^2 + \mu\mathrm{tr}[\varepsilon^2_-], \tag{7c}$$

where λ and μ are the Lamé's constants, and ε_+ and ε_- respectively identify the tensile and compressive counterparts of the strain tensor. In Equation (7), the symbol $\mathrm{tr}[\bullet]$ represents the trace operator, while $\langle \bullet \rangle_\pm$ stands for the so-called Macaulay bracket: $\langle \bullet \rangle_\pm = (\bullet \pm |\bullet|)/2$.

Finally, the degradation function $\mathfrak{g}(\eth)$, which introduces a stiffness deterioration with the growth of the defect (crack), can be defined as:

$$\mathfrak{g}(\eth) = (1 - \eth)^2 + \mathcal{K}, \tag{8}$$

where $\mathcal{K}$ is a residual stiffness parameter to prevent numerical instability issues for fully degraded states, i.e., when $\eth \approx 1$. Also, note that according to Equation (7a), the degradation function (Equation (8)) exclusively affects the positive part of the elastic energy of the body. Thus, the corresponding Cauchy stress tensor in a decomposed fashion for the PF formulation renders:

$$\sigma := \frac{\partial \hat{\psi}}{\partial \varepsilon} = \mathfrak{g}(\eth)\sigma_+ + \sigma_-; \quad \text{where } \sigma_\pm = \lambda\left(\langle \mathrm{tr}[\varepsilon]\rangle_\pm\right)\mathbf{1} + 2\mu\varepsilon_\pm. \tag{9}$$

In the Equation (9), **1** represents the second-order identity matrix and $\sigma_\pm$ identify the positive and negative parts of the stress tensor.

2.2. Numerical Implementation

In this Section, the numerical implementation of the PF approach for the fracture of intact rock, within the context of nonlinear FEM are briefly addressed. In the sequel, the corresponding to bulk operators will be denoted by the label *b* in the superscript.

Based on the isoparametric interpolation, the LaGrangian shape functions at the element level $N^I(\boldsymbol{\xi})$, defined in the natural space $\boldsymbol{\xi} = \{\xi^1, \xi^2\}$, are used for the interpolation of the geometry (**x**). The displacement field (**u**), as well as the variation ($\delta\mathbf{u}$) and the linearization ($\Delta\mathbf{u}$) of **u** are defined as:

$$\mathbf{x} \cong \textstyle\sum_{I=1}^n N^I\tilde{\mathbf{x}}_I = \mathbf{N}\tilde{\mathbf{x}}; \quad \mathbf{u} \cong \textstyle\sum_{I=1}^n N^I\mathbf{d}_I = \mathbf{Nd}; \quad \delta\mathbf{u} \cong \textstyle\sum_{I=1}^n N^I\delta\mathbf{d}_I = \mathbf{N}\delta\mathbf{d}; \quad \Delta\mathbf{u} \cong \textstyle\sum_{I=1}^n N^I\Delta\mathbf{d}_I = \mathbf{N}\delta\mathbf{d}, \tag{10}$$

where n indicates the number of nodes at the element level, and $\mathbf{x}_I$ y $\mathbf{d}_I$ denotes the nodal coordinates and displacements, respectively, which are collected into the operator $\tilde{\mathbf{x}}$ y $\mathbf{d}$. Accordingly, the interpolation functions are arranged into the vector form **N**.

In the similar lines, the strain field (ε), its variation ($\delta\varepsilon$) and its linearization ($\Delta\varepsilon$) can be approximated using the compatibility operator $\mathbf{B_d}$ associated with the kinematic field:

$$\varepsilon \cong \mathbf{B_d d}; \quad \delta\varepsilon \cong \mathbf{B_d}\delta\mathbf{d}; \quad \Delta\varepsilon \cong \mathbf{B_d}\Delta\mathbf{d}. \tag{11}$$

Similarly, the interpolation of the PF variable (damage-like variable), its variation and linearization render:

$$\eth \cong \sum_{I=1}^n N^I\overline{\eth}_I = \mathbf{N}\overline{\eth}; \quad \delta\eth \cong \sum_{I=1}^n N^I\delta\overline{\eth}_I = \mathbf{N}\delta\overline{\eth}; \quad \Delta\eth \cong \sum_{I=1}^n N^I\Delta\overline{\eth}_I = \mathbf{N}\delta\overline{\eth}, \tag{12}$$

where $\bar{\mathfrak{d}}_I$ are the nodal values of the PF variable, which are arranged into the vector $\bar{\mathfrak{d}}$. The spatial gradient of the PF ($\nabla_x \mathfrak{d}$), its variation ($\nabla_x \delta \mathfrak{d}$) and its linearization ($\nabla_x \Delta \mathfrak{d}$) can be interpolated through the operator $\mathbf{B}_{\mathfrak{d}}$:

$$\nabla_x \mathfrak{d} \cong \mathbf{B}_{\mathfrak{d}}\bar{\mathfrak{d}}; \quad \nabla_x(\delta\mathfrak{d}) \cong \mathbf{B}_{\mathfrak{d}}\delta\bar{\mathfrak{d}}; \quad \nabla_x(\Delta\mathfrak{d}) \cong \mathbf{B}_{\mathfrak{d}}\Delta\bar{\mathfrak{d}}. \tag{13}$$

Therefore, the discrete version of Equation (6) at the element level, identified by the super-index el, can be expressed as:

$$\delta\tilde{\Pi}_b^{el}(\mathbf{d}, \delta\mathbf{d}, \bar{\mathfrak{d}}, \delta\bar{\mathfrak{d}}) \tag{14}$$

$$= \delta\mathbf{d}^T \left\{ \int_{\Omega^{el}} \left[\left((1-\mathfrak{d})^2 + \mathcal{K}\right) \mathbf{B}_{\mathbf{d}}^T\sigma_+ + \mathbf{B}_{\mathbf{d}}^T\sigma_- \right] d\Omega - \int_{\partial\Omega^{el}} \mathbf{N}^T\bar{\mathbf{t}}\, d\partial\Omega - \int_{\Omega^{el}} \mathbf{N}^T\mathbf{f}_v\, d\Omega \right\}$$

$$+ \delta\bar{\mathfrak{d}}^T \left\{ \int_{\Omega^{el}} -2(1-\mathfrak{d})\mathbf{N}^T\psi_+^e(\varepsilon)\, d\Omega + \int_{\Omega^{el}} \mathcal{G}_c^b l \left(\mathbf{B}_{\mathfrak{d}}^T\nabla_x\mathfrak{d} + \frac{1}{l^2}\mathbf{N}^T\mathfrak{d} \right) d\Omega \right\}$$

$$= \delta\mathbf{d}^T\mathbf{f}_{\mathbf{d}}^b + \delta\bar{\mathfrak{d}}^T\mathbf{f}_{\mathfrak{d}}^b \tag{15}$$

where the residual vector $\mathbf{f}_{\mathbf{d}}^b = \mathbf{f}_{\mathbf{d},ext}^b - \mathbf{f}_{\mathbf{d},int}^b$, in which, $\mathbf{f}_{\mathbf{d},int}^b$ and $\mathbf{f}_{\mathbf{d},ext}^b$ represent the internal and external forces, respectively, given by:

$$\mathbf{f}_{\mathbf{d},int}^b = \int_{\Omega^{el}} \left[\left((1-\mathfrak{d})^2 + \mathcal{K}\right) \mathbf{B}_{\mathbf{d}}^T\sigma_+ + \mathbf{B}_{\mathbf{d}}^T\sigma_- \right] d\Omega, \tag{16}$$

$$\mathbf{f}_{\mathbf{d},ext}^b = \int_{\partial\Omega^{el}} \mathbf{N}^T\bar{\mathbf{t}}\, d\partial\Omega + \int_\Omega \mathbf{N}^T\mathbf{f}_v\, d\Omega. \tag{17}$$

Correspondingly, $\mathbf{f}_{\mathfrak{d}}^b$ is the residual vector associated with the PF variable, which can be estimated as:

$$\mathbf{f}_{\mathfrak{d}}^b = \int_{\Omega^{el}} -2(1-\mathfrak{d})\mathbf{N}^T\psi_+^e(\varepsilon)\, d\Omega + \int_{\Omega^{el}} \mathcal{G}_c^b l \left[\mathbf{B}_{\mathfrak{d}}^T\nabla_x\mathfrak{d}\frac{1}{l^2}\mathbf{N}^T\mathfrak{d} \right] d\Omega. \tag{18}$$

In this study, a staggered solution scheme in the form of Jacobi-type iterative procedure is adopted for the numerical treatment of the governing equations. Thus, the consistent linearization of Equation (14) yields the following coupled system [57]:

$$\begin{bmatrix} \mathbf{K}_{\mathbf{dd}}^b & 0 \\ 0 & \mathbf{K}_{\mathfrak{d}\mathfrak{d}}^b \end{bmatrix} \begin{bmatrix} \Delta\mathbf{d} \\ \Delta\mathfrak{d} \end{bmatrix} = \begin{bmatrix} \mathbf{f}_{\mathbf{d},ext}^b \\ 0 \end{bmatrix} - \begin{bmatrix} \mathbf{f}_{\mathbf{d},int}^b \\ \mathbf{f}_{\mathfrak{d}}^b \end{bmatrix}. \tag{19}$$

The particular forms of the matrices $\mathbf{K}_{\mathbf{dd}}^b$, $\mathbf{K}_{\mathbf{d}\mathfrak{d}}^b$, $\mathbf{K}_{\mathfrak{d}\mathbf{d}}^b$ and $\mathbf{K}_{\mathfrak{d}\mathfrak{d}}^b$ are omitted here for brevity [74]. The numerical implementation of the PF approach is carried out within the framework of the FE codes FEAP and ABAQUS, by generating the user-defined element routines.

3. Bulk Failure Predictions

3.1. Validation of the Developed PF Approach for Fracture Using the Brazilian Splitting Test

The developed PF methodology for fracture of intact rocks is validated by comparing the results with the standard Brazilian splitting test reported in [64]. The Brazilian test is extensively employed to characterize the tensile strength of rocks. The geometric definition of the benchmark problem involves a specimen of diameter equal to 52 mm, see Figure 2a. Furthermore, the following material parameters are considered: Young's modulus $E = 31.5$ GPa, Poisson's ratio $\nu = 0.25$ and $\mathcal{G}_c^b = 100$ J/m^2. Furthermore, the length scale in the PF approach can be related to the material strength σ_c of one-dimensional bar:

$$l = \frac{27E\mathcal{G}_c^b}{256\sigma_c^2}. \tag{20}$$

Therefore, in the current application, we set $l = 1$ mm. Half of the specimen is discretized using 55,452 elements, the simulations are performed by controlling the displacement. Figure 2b depicts the crack path estimated by implementing the present PF method in the ABAQUS platform, in line with [64]. During the simulations, it is observed that the crack initiated at the center of the disc, which agrees with the experimental evidence. Upon further loading, the initiated crack starts to grow towards the upper end of the Brazilian disc. From the qualitative standpoint, Figure 3 shows the experimental-numerical correlation, whereby a very satisfactory agreement can be observed along the loading path, as compared to the PF methods in [64]. In particular, the simulation response is characterized by a linear evolution until the abrupt failure and hence the fracture. Deviation of the simulated results from the experimental data is attributed to the absence of contact modeling features in the simulations.

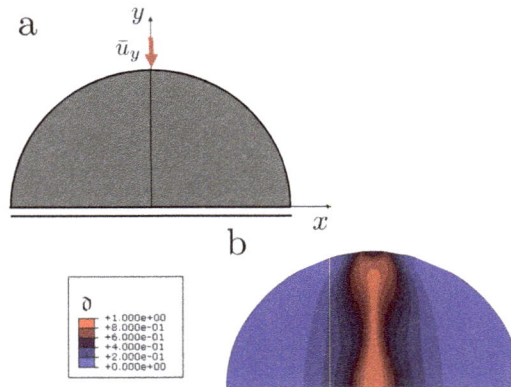

Figure 2. (**a**) Brazilian splitting test: loading condition. (**b**) Damage pattern of the specimen.

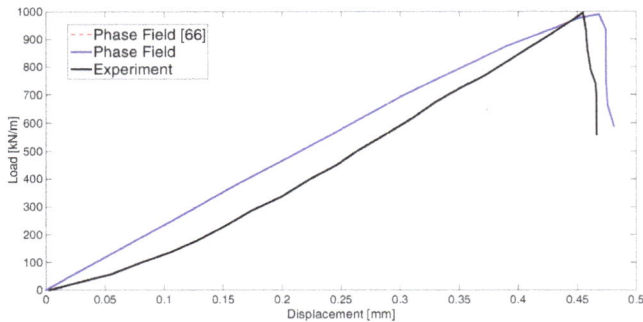

Figure 3. Brazilian splitting test: Load-displacement evolution curve.

3.2. Numerical Predictions of Rock Fracture: Estimation of Damage Patterns Various Load Cases for Specimens with Multiple Flaws

In this section, several examples are presented to illustrate the potential of the developed PF approach to simulate fracture phenomena in rock specimens, with an inclined echelon flaws (with respect to the horizontal direction), subjected to different load cases.

The basic geometry under consideration is a rock specimen with dimensions 200 mm × 200 mm, consisting of four inclined echelon flaws which can be simplified due to symmetry, to a quarter specimen of size 50 mm × 50 mm with a single inclined flaw, as shown in Figure 4. The adopted mechanical properties in the simulation are: $E = 30$ GPa, $\nu = 0.3$, $\mathcal{G}_c = 3$ J/m^2 and $l = 0.1$ mm [64].

The domain is discretized using first-order iso-parametric elements equipped with PF attributes, and a uniform mesh of size $h = 0.2$ mm. The specimen is subjected to displacements imposed along the vertical and horizontal directions, see Figure 4. In particular, the maximum vertical displacement is prescribed as $\bar{u}_y = \pm 3 \times 10^{-2}$ mm (tensile/compression), whereas $\bar{u}_x$ can have different ratios with respect to $\bar{u}_y$ depending on the specific case. Simulations are performed under controlled displacement, with a maximum displacement increment equal to $\Delta u = 1 \times 10^{-5}$ mm using a monolithic PF formulation.

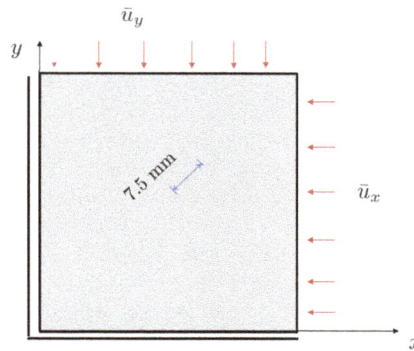

Figure 4. Specimen with inclined flaws: geometry and loading conditions.

3.2.1. Uniaxial Compression Test

We first analyze the deformation patterns in a uniaxial compression test, which is achieved by specifying $\bar{u}_y = -3 \times 10^{-2}$ mm, in several steps and $\bar{u}_x = 0$, see Figure 4. Figure 5 shows the failure maps, highlighting the fracture patterns, when the specified displacement reaches $\bar{u}_y = -2.4 \times 10^{-2}$ mm (left) and $\bar{u}_y = -2.78 \times 10^{-2}$ mm (right), respectively. Moreover, according to Figure 5, two cracks are observed to first initiate around the tips of the flaw, which are subsequently propagated parallel to the loading direction. These predictions are in good agreement with experimental observations, see [64].

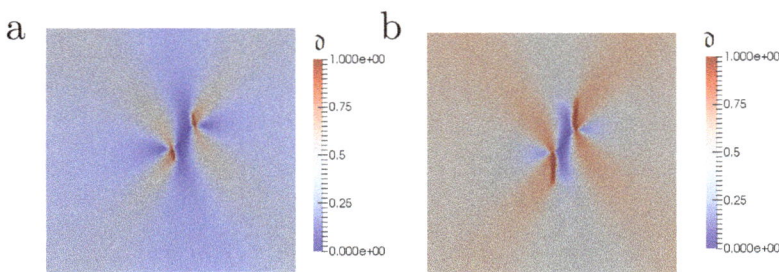

Figure 5. Uniaxial compression case: Failure pattern at an initial stage (**a**) and at advanced stage (**b**).

3.2.2. Uniaxial Tensile Test

In the next step, the fracture patterns in a uniaxial tensile test are captured using the developed PF approach. Therefore, a tensile load along the y direction equal to $\bar{u}_y = 3 \times 10^{-2}$ mm, is specified on to the system shown in Figure 4. The simulation is carried out by specifying the displacement load in several steps. Figure 6a. shows the failure maps highlighting the fracture patterns, when the specified displacement reaches $\bar{u}_y = -2.4 \times 10^{-2}$ mm (left) and $\bar{u}_y = -2.78 \times 10^{-2}$ mm (right), respectively.

As compared to Figure 5, the predicted crack pattern is perpendicular to the loading direction, which agrees with the results reported in [64].

To summarize, the estimated results from compression and tension tests, using the present PF method are in close agreement with the reported experimental results [64], illustrating the robustness of the proposed methodology.

3.2.3. Compressive-Traction and Traction-Traction Tests

To complement the previous results, the developed methodology is extended to predict the fracture patterns under compression-traction and traction-traction load cases. The particular load cases are achieved by suitably combining the imposed vertical and horizontal displacements.

Figure 6b,c depict the failure maps under uniaxial tensile load, vertical compression and horizontal traction with $\bar{u}_y/\bar{u}_x = -1$ and vertical compression and horizontal traction cases, respectively, setting $\bar{u}_y = -2.5 \times 10^{-2}$ mm. Notable incidence of the external loadings with respect to the estimated crack patterns can be observed. Figure 6b,c correspond to the failure maps of the samples subjected to compression-traction loads with $\bar{u}_y/\bar{u}_x = -1$ and $\bar{u}_y/\bar{u}_x = -2$, respectively. Comparing Figure 6b,c, the applied horizontal displacement is observed to have a moderate influence on the final damage pattern, which is principally aligned with the vertical imposed displacement.

The analysis is also applied to samples subjected to tensile-tensile loads. Figure 7a–c correspond to the failure maps estimated by specifying the displacement ratios $\bar{u}_y/\bar{u}_x = 1, 2$, and 10, respectively. We note that the predicted crack patterns evolve from a perfectly aligned path with the flaw orientation for $\bar{u}_y/\bar{u}_x = 1$, to almost horizontal path for $\bar{u}_y/\bar{u}_x = 10$. In line with previous cases, the estimated results agree with the reported results in [64].

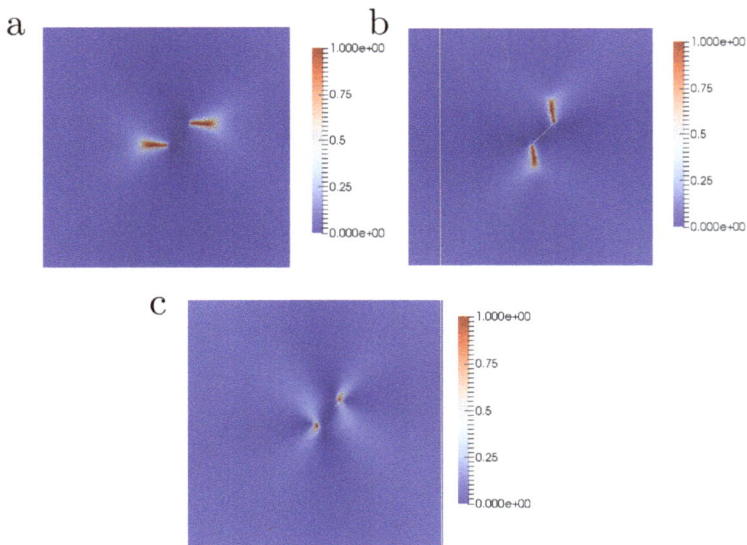

Figure 6. Different loading scenarios for a specimen with a single flaw: predicted damage patterns. (**a**) Uniaxial tensile case. (**b**) Vertical compression and horizontal traction with $\bar{u}_y/\bar{u}_x = -1$. (**c**) Vertical compression and horizontal traction with $\bar{u}_y/\bar{u}_x = -2$.

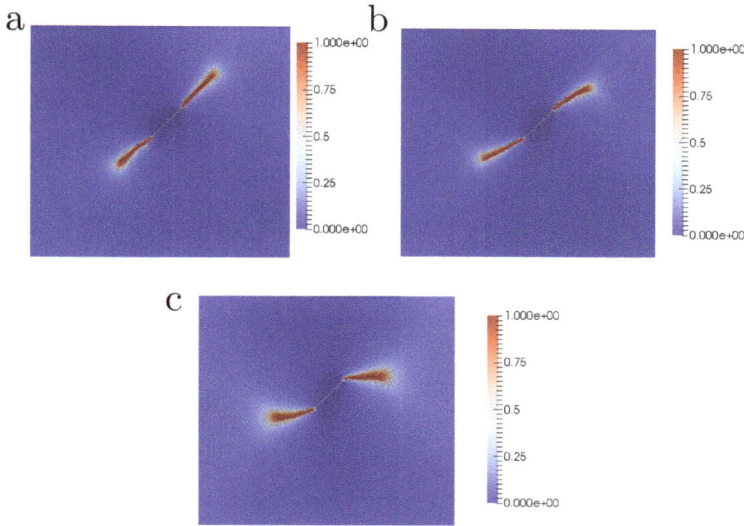

Figure 7. Different tensile-tensile loading scenarios for a specimen with a single flaw: predicted damage patterns. (**a**) Vertical traction and horizontal traction with $\bar{u}_y/\bar{u}_x = 1$. (**b**) Vertical traction and horizontal traction with $\bar{u}_y/\bar{u}_x = 2$. (**c**) Vertical traction and horizontal traction with $\bar{u}_y/\bar{u}_x = 10$.

4. Coupling the PF Approach for Rock Fracture and the Interface Cohesive Zone Model: Application to Heterogeneous Media

4.1. General Aspects

In many engineering applications, rock mechanics in particular, heterogenous media separated by existing interfaces are often observed. The simulation of fracture events in such media is a challenging task. In this context, a plausible alternative is the combined phase field-cohesive zone model (PF-CZM) numerical approach [68], where a consistent coupling between the PF approach for brittle fracture in the matrix and the interface crack model are proposed. This technique can be conceived as a new paradigm for the consistent coupling of damage events of different signature, allowing complex crack evolution with multiple branches.

The point of departure of the PF-CZM consists of the consideration of a system with a predefined crack Γ_b and a prescribed interface Γ_i, see Figure 8. A generic point of the interface is denoted by the vector $\mathbf{x}_c$. Considering the internal potential of the body (omitting the subscript *int* to alleviate the notation), the governing functional of the system (Equation (6)) can be written as:

$$\Pi(\mathbf{u}, \Gamma) = \Pi_\Omega(\mathbf{u}, \Gamma) + \Pi_\Gamma(\Gamma) = \int_{\Omega \backslash \Gamma} \psi^e(\boldsymbol{\varepsilon}) \, \mathrm{d}\Omega + \int_\Gamma \mathcal{G}_c \, \mathrm{d}\Gamma, \tag{21}$$

The main attribute of the current method renders the additive decomposition of the dissipative part of the functional $\mathcal{G}_c$ into two components related to: (*i*) the fracture energy dissipated in the matrix rock, which is characterized by $\mathcal{G}_c^b$ and is modeled using the PF method, and (*ii*) the interface energy $\mathcal{G}^i$ that identifies the crack propagation along the existing interfaces and is modeled using a cohesive zone representation. Correspondingly, the governing functional in Equation (21), can be rewritten as:

$$\Pi(\mathbf{u}, \Gamma_b, \Gamma_i) = \Pi_\Omega + \Pi_{\Gamma_b} + \Pi_{\Gamma_i} = \int_{\Omega \backslash \Gamma} \psi^e(\boldsymbol{\varepsilon}) \, \mathrm{d}\Omega + \int_{\Gamma_b} \mathcal{G}_c^b(\mathbf{u}, \partial) \, \mathrm{d}\Gamma + \int_{\Gamma_i} \mathcal{G}^i(\mathbf{g}, \hbar, \partial) \, \mathrm{d}\Gamma, \tag{22}$$

where **g** is used to identify the displacement gaps between the two flanks of the interface, $\mathfrak{h}$ is a history parameter [58], and $\mathfrak{d}$ is the PF variable in the matrix.

In this work, we account for the tension cut-off interface behavior given in [68], whose apparent stiffness is modified by taking into consideration the matrix damage state in the surrounding region via $\mathfrak{d}$. Moreover, it is worth mentioning that the interface fracture energy, Equation (22), can be decomposed into the corresponding contributions to the fracture Modes I and II in a 2D setting of analysis, which are represented by $\mathcal{G}_{IC}^i$ and $\mathcal{G}_{IIC}^i$, respectively.

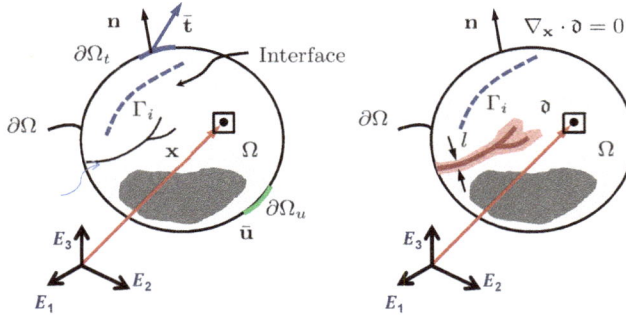

Figure 8. Coexistence between brittle fracture in the bulk and cohesive debonding of an interface within the context of the phase field approach of fracture.

According to [68], it is assumed that the critical interface opening (g_c) obeys a linear law with respect to the damage state in the matrix. Therefore, the following relation can be postulated: $g_c(\mathfrak{d}) = (1 - \mathfrak{d})g_{c,0} + \mathfrak{d}g_{c,1}$, where $g_{c,0} = g_c(\mathfrak{d} = 0)$ and $g_{c,1} = g_c(\mathfrak{d} = 1)$. Then the traction-separation law at the interface (Figure 9 for mode I), are given by:

$$\sigma = \begin{cases} k_n \dfrac{g_n}{g_{nc}}, & \text{if } 0 < \dfrac{g_n}{g_{nc}} < 1; \\ 0, & \text{if } \dfrac{g_n}{g_{nc}} \geq 1, \end{cases} \tag{23}$$

$$\tau = \begin{cases} k_t \dfrac{g_t}{g_{tc}}, & \text{if } 0 < \dfrac{g_t}{g_{tc}} < 1; \\ 0, & \text{if } \dfrac{g_t}{g_{tc}} \geq 1, \end{cases} \tag{24}$$

where σ and τ are the critical traction values associated with the fracture Modes I and II, respectively, g are the displacements whose sub-indices n and t refer to fracture Modes I and II, respectively.

Consequently, the apparent interface stiffnesses for both fracture modes as a function of the bulk damage $\mathfrak{d}$ render

$$k_n = k_{n,0} \left(\frac{g_{nc,0}}{g_{nc}} \right)^2; \quad k_t = k_{t,0} \left(\frac{g_{tc,0}}{g_{tc}} \right)^2, \tag{25}$$

where k_0 and g_0 represent for the interface stiffness corresponding to intact matrix rock, i.e., $\mathfrak{d} = 0$.

We adopt a quadratic-based fracture criterion for triggering propagation events, given by:

$$\left(\frac{\mathcal{G}_I^i}{\mathcal{G}_{IC}^i} \right)^2 + \left(\frac{\mathcal{G}_{II}^i}{\mathcal{G}_{IIC}^i} \right)^2 = 1, \tag{26}$$

where $\mathcal{G}_I^i$ and $\mathcal{G}_{II}^i$ are the individual energy release rates (ERRs) for fracture Modes I and II, respectively, which take the form:

$$\mathcal{G}_I^i(\eth) = \frac{1}{2}n_{t,0}g_n^2 \frac{g_{nc,0}^2}{[(1-\eth)g_{nc,0} + \eth g_{nc,1}]^2}; \quad \mathcal{G}_{II}^i(\eth) = \frac{1}{2}k_{t,0}g_t^2 \frac{g_{tc,0}^2}{[(1-\eth)g_{tc,0} + \eth g_{tc,1}]^2}. \tag{27}$$

Critical values of the ERRs defined above are compared to their critical values via Equation (26), whereby $\mathcal{G}_{IC}^i$ and $\mathcal{G}_{IIC}^i$ are the critical values, given by:

$$\mathcal{G}_{IC}^i = \frac{1}{2}g_{nc,0}^2 k_{n,0}; \quad \mathcal{G}_{IIC}^i = \frac{1}{2}g_{tc,0}^2 k_{t,0}. \tag{28}$$

Finally, the interface formulation to encompass geo-mechanical aspects can be arrived in the similar lines of [75], where we incorporate a modified version of the normal penalty stiffness (to prevent inter-penetration between the interface flanks), which obeys the following hyperbolic expression:

$$g_n \leq 0 \quad \rightarrow \quad \sigma = \frac{k_{n,0}g_n}{1 - g_n/g_m}, \tag{29}$$

where g_n is the interface closure, $k_{n,0}$ is the initial stiffness, g_m is the maximum interface closure. The values $k_{n,0}$ and g_m, can be related in rock mechanics to the joint compressive strength (JCS), which is provided in the literature for many different geological masses, using the following correlations (Figure 9):

$$k_{n,0} = -7.15 + 1.75\,\text{JRC} + 0.02\left(\frac{\text{JCS}}{a_0}\right)^{-0.2510} \tag{30}$$

$$g_m = -7.15 + 1.75\,\text{JRC} + 0.02\left(\frac{\text{JCS}}{a_0}\right)^{-0.2510} \tag{31}$$

where JCS is the joint wall compression strength, which is the joint roughness coefficient introduced by Barton and Choubey [76], see also a recent overview on the use of this index in [77] and $a_0 = \text{JRC}_0/50$. These coefficients can be characterized by means of experimental campaigns on specific rock masses.

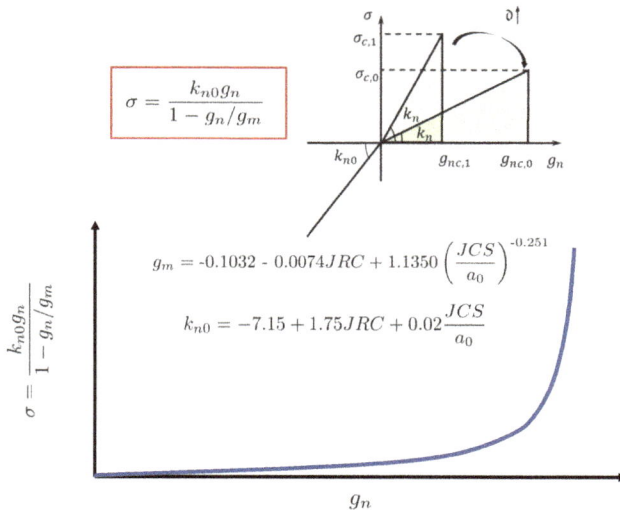

Figure 9. Modification of interface penalty stiffness under normal compression due to geo-mechanical aspects: schematic representation under negative normal gap g_n.

In line with the description of the current model, it is important to note that the independent parameters that characterize the interface response are: (*i*) the critical normal and tangential displacements corresponding to Mode I and II of fracture g_n and g_t, (*ii*) the associated interface fracture energies and the fracture criterion, and (*iii*) the corresponding geo-mechanical parameters. Additionally, the present model includes an additional capability that couples the apparent interface stiffness with respect to the PF (bulk) damage variable, and therefore two extra parameters are required corresponding to the pairs: $g_{n,0}$, g_n and $g_{t,0}$, g_t.

4.2. Variational Form and Finite Element Formulation

In this section, we present the particular aspects concerning the interface contribution to the total potential of the system, Equation (22). Following a standard Galerkin procedure, the weak form of the interface contribution can be expressed as:

$$\delta \Pi_{\Gamma_i}(\mathbf{u}, \delta \mathbf{u}, \eth, \delta \eth) = \int_{\Gamma_i} \left(\frac{\partial \mathcal{G}^i(\mathbf{u}, \eth)}{\partial \mathbf{u}} \delta \mathbf{u} + \frac{\partial \mathcal{G}^i(\mathbf{u}, \eth)}{\partial \eth} \delta \eth \right) d\Gamma, \tag{32}$$

$$\delta \Pi_{\Gamma_i}(\mathbf{u}, \delta \mathbf{u}, \eth, \delta \eth) = \int_{\Gamma_i} \left(\frac{\partial \mathcal{G}^i(\mathbf{u}, \eth)}{\partial \mathbf{u}} \delta \mathbf{u} + \frac{\partial \mathcal{G}^i(\mathbf{u}, \eth)}{\partial \eth} \delta \eth \right) d\Gamma, \tag{33}$$

where $\delta \mathbf{u}$ are the test functions of the displacement field $(\mathfrak{V}^u = \{\delta \mathbf{u} \,|\, \mathbf{u} = \overline{\mathbf{u}} \text{ on } \partial \Omega_u, \mathbf{u} \in \mathcal{H}^1\})$, and $\delta \eth$ stands for the test functions of the PF variable $d (\mathfrak{V}^\eth = \{\delta \eth \,|\, \delta \eth = 0 \text{ on } \Gamma_b, \eth \in \mathcal{H}^0\})$.

The discretization of the present interface formulation is performed using first-order elements with linear interpolation functions. Thus, analogous to the PF approach for bulk fracture presented in Section 2.2, **d** denotes the nodal displacement vector and $\bar{\eth}$ is the phase field nodal vector, both at the element level. Accordingly, the discrete form of Equation (32) at the element level Γ_i^{el} $(\Gamma_i \sim \bigcup \Gamma_i^{el})$ renders:

$$\delta \tilde{\Pi}_{\Gamma_i}^{el}(\mathbf{d}, \delta \mathbf{d}, \bar{\eth}, \delta \bar{\eth}) = \int_{\Gamma_i^{el}} \left(\frac{\partial \mathcal{G}^i(\mathbf{d}, \bar{\eth})}{\partial \mathbf{d}} \delta \mathbf{d} + \frac{\partial \mathcal{G}^i(\mathbf{d}, \bar{\eth})}{\partial \bar{\eth}} \delta \bar{\eth} \right) d\Gamma. \tag{34}$$

The vector of displacement gaps (**g**) for any arbitrary point on the surface Γ_i^{el} is the result of the difference in displacement vector between opposite flanks, which can be obtained from the displacement field **d** by performing a pre-multiplication by the operator **L**:

$$\mathbf{g} = \mathbf{NLd} = \hat{\mathbf{B}}_\mathbf{d}\mathbf{d}, \tag{35}$$

where **N** is a matrix including the element shape functions of the displacement field and $\hat{\mathbf{B}}_\mathbf{d} = \mathbf{NL}$ stands for the compatibility operator of the displacement field associated with the interface.

The transformation of the global gap vector into the local setting referred to the interface [38,41] is conducted via the rotation matrix **R** from Equation (35) leading to the local gap vector $\mathbf{g}_{loc}$, given by:

$$\mathbf{g}_{loc} \cong \mathbf{Rg} = \mathbf{R}\hat{\mathbf{B}}_\mathbf{b}\mathbf{d}. \tag{36}$$

In a similar manner, the following discretization forms can be deduced for the average PF variable $\eth$ at the interface, at the element level Γ_i^{el} as:

$$\eth \cong \mathbf{N}_\eth \mathbf{M}_\eth \bar{\eth} = \hat{\mathbf{B}}_\eth \bar{\eth}, \tag{37}$$

where $\mathbf{M}_\eth$ is an operator computing the average value of the PF variable between the two interface flanks and $\hat{\mathbf{B}}_\eth = \mathbf{N}_\eth \mathbf{M}_\eth$ identifies the corresponding compatibility operator. The specific forms of these matrices are given in [39,41].

Therefore, the discrete weak form renders

$$
\begin{aligned}
\delta \tilde{\Pi}_{\Gamma_i}^{el}(\mathbf{d}, \delta \mathbf{d}, \eth, \delta \eth) &= \delta \mathbf{d}^{\mathrm{T}} \int_{\Gamma_i^{el}} \left(\frac{\partial \mathcal{G}^i(\mathbf{d}, \eth)}{\partial \mathbf{d}} \right)^{\mathrm{T}} d\Gamma + \delta \eth^{\mathrm{T}} \int_{\Gamma_i^{el}} \left(\frac{\partial \mathcal{G}^i(\mathbf{d}, \eth)}{\partial \eth} \right)^{\mathrm{T}} d\Gamma \\
&= \delta \mathbf{d}^{\mathrm{T}} \int_{\Gamma_i^{el}} \hat{\mathbf{B}}_\mathbf{d}^{\mathrm{T}} \mathbf{R}^{\mathrm{T}} \left(\frac{\partial \mathcal{G}^i(\mathbf{d}, \eth)}{\partial \mathbf{g}_{loc}} \right)^{\mathrm{T}} d\Gamma + \delta \eth^{\mathrm{T}} \int_{\Gamma_i^{el}} \hat{\mathbf{B}}_\eth^{\mathrm{T}} \left(\frac{\partial \mathcal{G}^i(\mathbf{d}, \eth)}{\partial \eth} \right)^{\mathrm{T}} d\Gamma,
\end{aligned}
\tag{38}
$$

leading to the residual vectors:

$$
\mathbf{f}_\mathbf{d}^i = \int_{\Gamma_i^{el}} \hat{\mathbf{B}}_\mathbf{d}^{\mathrm{T}} \mathbf{R}^{\mathrm{T}} \left(\frac{\partial \mathcal{G}^i(\mathbf{d}, \eth)}{\partial \mathbf{g}_{loc}} \right)^{\mathrm{T}} d\Gamma,
\tag{39a}
$$

$$
\mathbf{f}_\eth^i = \int_{\Gamma_i^{el}} \hat{\mathbf{B}}_\eth^{\mathrm{T}} \left(\frac{\partial \mathcal{G}^i(\mathbf{d}, \eth)}{\partial \eth} \right)^{\mathrm{T}} d\Gamma.
\tag{39b}
$$

Finally, through the directional derivative, the linearization of the residual vectors yields the following tangent matrices:

$$
\mathbf{K}_{\mathbf{dd}}^i = \frac{\partial \mathbf{f}_\mathbf{d}}{\partial \mathbf{d}} = \int_{\Gamma_i^{el}} \hat{\mathbf{B}}_\mathbf{d}^{\mathrm{T}} \mathbf{R}^{\mathrm{T}} \mathbb{C}_{\mathbf{dd}}^i \mathbf{R} \hat{\mathbf{B}}_\mathbf{d} \, d\Gamma,
\tag{40a}
$$

$$
\mathbf{K}_{\mathbf{d}\eth}^i = \frac{\partial \mathbf{f}_\mathbf{d}}{\partial \eth} = \int_{\Gamma_i^{el}} \hat{\mathbf{B}}_\mathbf{d}^{\mathrm{T}} \mathbf{R}^{\mathrm{T}} \mathbb{C}_{\mathbf{d}\eth}^i \hat{\mathbf{B}}_\eth \, d\Gamma,
\tag{40b}
$$

$$
\mathbf{K}_{\eth\mathbf{d}}^i = \frac{\partial \mathbf{f}_\eth}{\partial \mathbf{d}} = \int_{\Gamma_i^{el}} \hat{\mathbf{B}}_\eth^{\mathrm{T}} \mathbb{C}_{\eth\mathbf{d}}^i \mathbf{R} \hat{\mathbf{B}}_\mathbf{d} \, d\Gamma,
\tag{40c}
$$

$$
\mathbf{K}_{\eth\eth}^i = \frac{\partial \mathbf{f}_\eth}{\partial \eth} = \int_{\Gamma_i^{el}} \hat{\mathbf{B}}_\eth^{\mathrm{T}} \mathbb{C}_{\eth\eth}^i \hat{\mathbf{B}}_\eth \, d\Gamma,
\tag{40d}
$$

where:

$$
\mathbb{C}_{\mathbf{dd}}^i = \begin{bmatrix} \hat{\alpha} k_n & 0 \\ 0 & \hat{\beta} k_t \end{bmatrix},
\tag{41a}
$$

$$
\mathbb{C}_{\mathbf{d}\eth}^i = \begin{bmatrix} g_n k_n \dfrac{\partial \hat{\alpha}}{\partial \eth}, & g_t k_t \dfrac{\partial \hat{\beta}}{\partial \eth} \end{bmatrix},
\tag{41b}
$$

$$
\mathbb{C}_{\eth\mathbf{d}}^i = \begin{bmatrix} g_n k_n \dfrac{\partial \hat{\alpha}}{\partial \eth} \\ g_t k_t \dfrac{\partial \hat{\beta}}{\partial \eth} \end{bmatrix},
\tag{41c}
$$

$$
\mathbb{C}_{\eth\eth}^i = \frac{1}{2} g_n^2 k_n \frac{\partial^2 \hat{\alpha}}{\partial \eth^2} + \frac{1}{2} g_t^2 k_t \frac{\partial^2 \hat{\beta}}{\partial \eth^2},
\tag{41d}
$$

and $\hat{\alpha}$ and $\hat{\beta}$ are given by:

$$
\hat{\alpha} = \frac{g_{nc,0}^2}{\left[(1 - \eth) g_{nc,0} + \eth g_{nc,1} \right]^2},
\tag{42a}
$$

$$
\hat{\beta} = \frac{g_{tc,0}^2}{\left[(1 - \eth) g_{tc,0} + \eth g_{tc,1} \right]^2}.
\tag{42b}
$$

Please note that in line with Equation (19), the resulting coupled equations are solved using the monolithic Newton-Raphson scheme, obeying the form:

$$\begin{bmatrix} \mathbf{K}_{dd}^i & \mathbf{K}_{d\eth}^i \\ \mathbf{K}_{\eth d}^i & \mathbf{K}_{\eth\eth}^i \end{bmatrix} \begin{bmatrix} \Delta\mathbf{d} \\ \Delta\eth \end{bmatrix} = \begin{bmatrix} \mathbf{f}_d^i \\ \mathbf{f}_{\eth}^i \end{bmatrix}. \tag{43}$$

5. Fracture Analysis in Heterogeneous Rock Masses Using the PF-CZM Approach

5.1. General Considerations

The presence of interfaces in joint propagation is especially relevant in sedimentary rocks composed of siltstone and shale layers. The analysis of cracking events in multilayered materials has been a recurrent research in the last few years, as described in [68]. Similar to composite materials, there are three fundamental aspects of crucial importance in fracturing across interfaces, called joints in the rock mechanics literature: (*i*) the strength of the joint, (*ii*) the material properties of the layers and (*iii*) the loading conditions.

As discussed in [78], when an approaching crack impinges on a tough interface this crack generally tends to propagate into the adjacent bulk without experience significant deviations from its initial path. Conversely, weak joints are much more prone to fail, leading to noticeable delamination events. This basic scenario can be reproduced in a straightforward manner by the PF-CZM approach outlined in the previous section. Thus, Figure 10 qualitatively depicts the simulation of a homogeneous system of span $L = 50$ mm with the mechanical properties given above with a prescribed horizontal interface, which is then subjected to a uniform displacement at the lateral edges $\bar{u}_x$. We set the ratio between the fracture toughness of the bulk and the interface as $\mathcal{G}_c^b / \mathcal{G}_c^i = 50$ for weak joint scenario, whereas, $\mathcal{G}_c^b / \mathcal{G}_c^i = 0.05$ for tough joints. In this graph, it can be observed that the proposed PF-CZM faithfully reproduces the basic behavior discussed in [78]. Specifically, a discontinuous displacement map is shown for the weak joint case illustrating delamination events, whereas a well-defined crack front perpendicular to the loading direction is reproduced for the strong joint response.

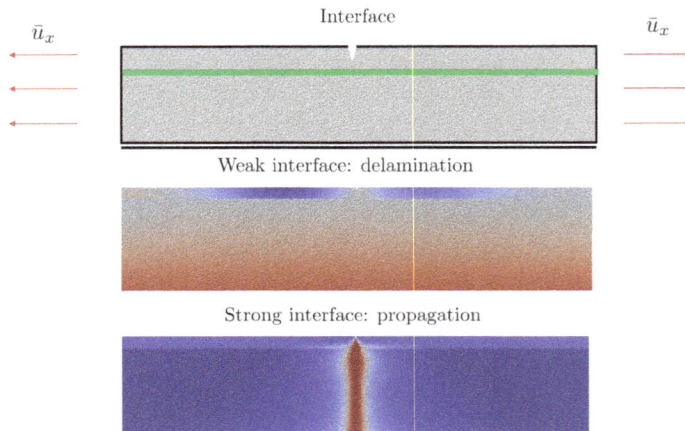

Figure 10. String system subjected to uniaxial displacement. Predictions using PF-CZM for two types of interfaces between similar materials: weak joints with delamination events along the interface and strong joints with continuous crack path.

However, for bi-material systems with weak joints, more complex situations can be encountered according to LEFM studies [68]. Particularly, three different propagation scenarios can be found based on the specific Dundurs's parameters of the system when a crack impinges on an interface: (1) single deflection along the interface, (2) double deflection along the interface and (3) penetration. Figure 11 succinctly reproduces the results discussed in [68] using the current PF-CZ approach for a bi-material system subjected to uniform tensile load with a prescribed vertical joint between both media. Observing Figure 11, the numerical predictions led to very satisfactory qualitative agreement with respect to theoretical LEFM results in terms of fracture patterns for different values of the Dundurs's parameter α with respect to $\Pi_1 = \mathcal{G}_c^b / \mathcal{G}_c^i$.

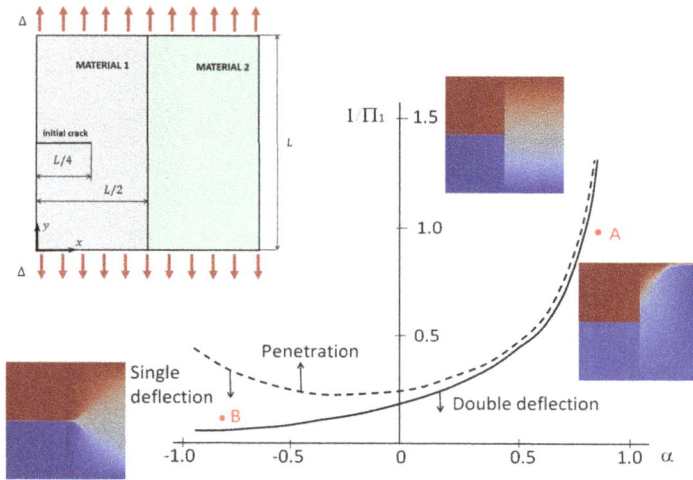

Figure 11. Geometry and boundary conditions for the bi-material problem. Crack pattern scenarios for brittle interface. The contour plots of the dimensionless vertical displacement field correspond to three different cases labeled A (double deflection), B (single deflection), C (penetration).

Relying on the illustrated potential capabilities, the next Sections addresses the applicability of the PF-CZ method to simulate fracture events in multilayered rock masses.

5.2. Application to Rock Salt Fracture with Inclined Interfaces

One of the potential applications of the developed numerical framework is the investigation of the response of bedded rocks salt. Particularly, these rocks perform in a different manner depending upon the corresponding lithology, which can obey either single or multi-layer arrangements. Following [79], mixed lithotypes present horizontal or inclined beds of anhydrite mudstone alternating with halite, leading to anhydrite-halite composite rocks.

From the experimental evidence in [79], it was observed that in such configurations, the presence of interlayers tends to promote crack initiation at this location, this fact being also promoted due to the inclined interface definition. Thus, these authors reported that typical failure sequence can be described by three different phases: (*i*) near/along the region of the interface, initial micro-defects stemming from the differences between sedimentation and consolidation evolve; (*ii*) increasing the applied loading, such micro-cracks propagate along the interface and within the rock salt region; (*iii*) finally, the coalescence of different crack occur, breaking the specimen.

The current numerical method is applied to study the deformation behavior and the fracture characteristics (initiation and coalescence) in bedded rocks salt with inclined interfaces under uniaxial compression in line with [79]. For this purpose, uniaxial compression simulations of composite rocks salt with interlayers of 20 mm in thickness are numerically analyzed, which showed a mechanical behavior close to brittle character. Due to the lack of reliable data regarding the geometry of the specimens analyzed in [79], the study is restricted to a qualitative investigation following a central interlayer disposal, see Figure 12. A domain of dimensions 100×200 mm is defined with a central interlayer of 20 mm in thickness following a macroscopic representation with two small flaws close to the interface. This domain is dsicretized using 84388 elements combining bulk and interface elements. The mechanical and fracture properties of both materials are listed in Table 1. Please note that for the interface between the rock salt and the anhydrite we choose an intermediate value between the transgranular boundary crack properties reported in [80], so that we set $\mathcal{G}_c^i = 1.2715$ J/m^2.

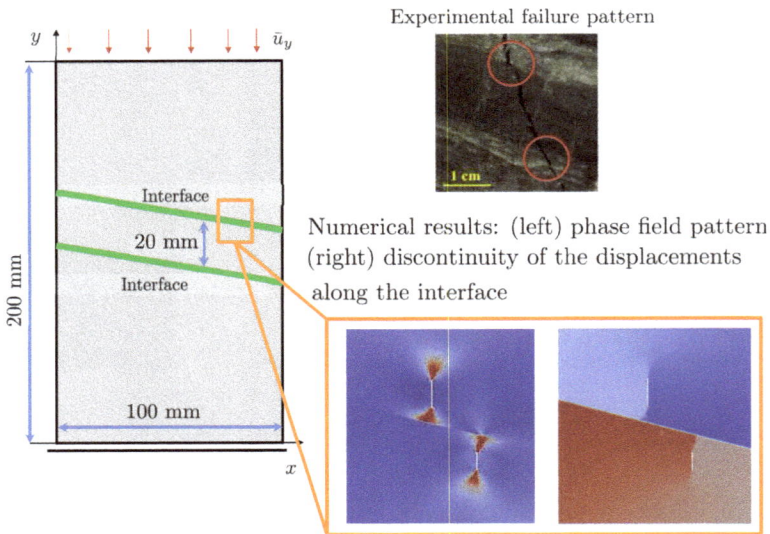

Figure 12. Salt rock composite specimen under uniaxial compressive loading. Geometry and boundary conditions, experimental failure pattern and numerical predictions.

Table 1. Mechanical and fracture properties: halite and anhydrite.

Material	E (GPa)	ν	$\mathcal{G}_c^b$ (Jm^{-2})
Halite	36.87	0.254	1.155
Anhydrite	74.36	0.269	1.805

Figure 12 depicts the qualitative numerical-experimental correlation of the predicted crack path of the current method and that observed in the experiments, showing a good qualitative agreement. In particular, during the simulations, it is observed that the main vertical cracks initiate at the flaws near the interface and propagate through the interlayer and subsequently both cracks propagate along the interface coalescing. The final failure pattern at this location agrees from a qualitative point of view with that reported in [79]. However note that the development of wing and shear crack paths are slightly deviated from the experimental evidence, so that alternative PF methods for bulk fracture as proposed in [65,66] can be employed in forthcoming studies. Nevertheless, a more comprehensive investigation falls beyond the scope of the current paper due to the lack of the required material data.

5.3. Numerical Investigation: Specimen with Multiple Flaws and Presence of Interfaces

Continuing with the assessment of the predictive capabilities of the current PF-CZM approach for heterogenous rock masses, we adopt the specimen definition described in Section 3, including the prescribed weak joints and multiple flaws with different lengths and orientations forming 45° with respect to the horizontal direction. Material properties of the specimens also replicate those given in Section 3. For the sake of conciseness in the results, we assume that the different lithologies are from the same material. Also note that bi- and tri-material systems can be directly analyzed using the current numerical methodology in a straightforward manner.

Additionally, with respect to the joint definitions that separate different bulk domains, we consider two cases: (*i*) a single horizontal joint, (*ii*) a single vertical joint. Joint mechanical properties are given in Table 2, whereas, contact stiffness properties relying on geo-mechanical parameters are listed in Table 3.

It is worth mentioning that due to the lack of reliable experimental data, after validating the proposed method in the previous Sections, the present results are conducted to analyze the potential fracture patterns for different geometric characteristics and loading conditions.

Table 2. Interface fracture properties.

Critical traction Mode I interface σ_{c0}	75 MPa
Critical traction Mode II interface τ_{c0}	90 MPa
Fracture toughness Mode I	0.002 N/mm
Fracture toughness Mode II	0.008 N/mm

Table 3. Penalty stiffness properties of the joint based on geo-mechanical aspects.

g_m	$10\, g_{n0}$
a_0	0.194 mm
JCS_0	169 MPa
JRC_0	9.7
Friction angle	32°
L_0	0.2 m

5.3.1. Specimen with Two Inclined Flaws and a Horizontal Joint

The specific description of the flaw definitions within the domain, the joint positions and the loading conditions are shown in Figure 13a. The position of the joint coincides with the medial horizontal edge of the specimen. A compression-tension load case, where $\bar{u}_y = -0.2\bar{u}_x$, with compression along the vertical direction is considered (we set $\bar{u}_y = 5 \times 10^{-2}$ mm in the following analysis here presented) Figure 13b,c depict the predicted contour plots of the vertical displacement and the damage pattern at the end of the simulation. Analyzing the displacement field, Figure 13b, a clear discontinuity at the position of the joint can be appreciated. This fact indicates the occurrence of joint failure and therefore delamination. However, almost no increase in the value of PF variable within the specimen is predicted, see Figure 13c. These results clearly indicate that delamination along the existing joint can be identified as the predominant failure phenomenon for the given loading conditions.

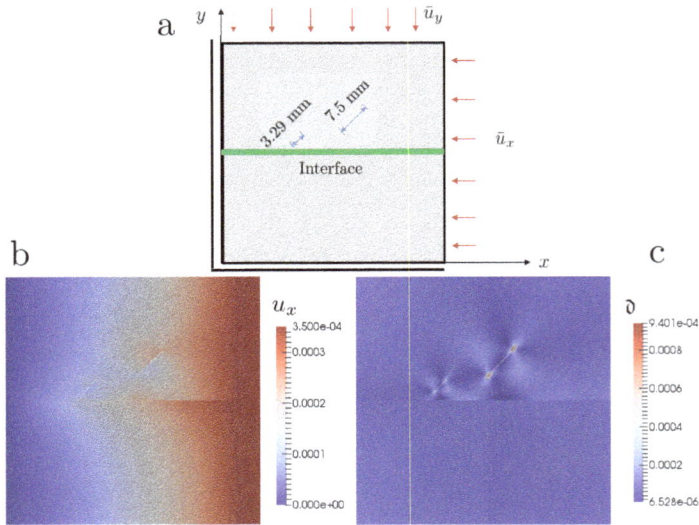

Figure 13. (**a**) Specimen with 2 inclined flaws and a horizontal joint. Case $\bar{u}_y = -0.2\bar{u}_x$: (**b**) horizontal displacement field; (**c**) damage pattern.

To examine the clear influence of the applied loading on the failure characteristics, we analyze the specimen by prescribing the traction-traction loading conditions, i.e., through the definition of $\bar{u}_y = 0.5\bar{u}_x$ (both positive values along the corresponding edges based on the Cartesian setting given in the graph). The computed results clearly pinpoint a substantial change in the fracture conditions within the specimen, see Figure 14. In this case, the development of nucleation and propagation phases around the flaw tips are noticed. Upon load increasing, both crack events emanating from the flaws coalesce and subsequently reach the joint. A third phase in the failure path evolution is manifested by the crack propagations along the existing joint and subsequently progressing into the bottommost substrate, which concomitantly occurs with the crack propagation in the topmost substrate. These phenomena can be identified by analyzing the discontinuities in the contour map associated with the vertical displacement field and the PF evolution. This concomitant simulation of joint and bulk damage can be identified as one of the most appealing attributes of the present numerical methodology, since this allows interaction without user intervention throughout the computations. The simultaneous evolution of failure events from different signature during the analysis can be identified in Figure 14 regarding the load-displacement curve, whereby a first linear elastic evolution is followed by crack growth from the two flaws with incipient delamination along the interface in a later stage. Failure initiation is around $u_y/\bar{u}_y = 0.53$.

Finally, for a most severe traction-traction case we define $\bar{u}_y = \bar{u}_x$. Analyzing the contour map corresponding to the vertical displacement, Figure 15a, it is observed that for this loading configuration the crack pattern tends to progress with gradual change in its path. In particular, bulk fracture is reoriented along the direction of the flaws, and then reduce the extent of the delamination front, as seen in Figure 15b. This effect is also noticed above for the homogeneous crack example with a single inclined flaw, in Section 5.3. This behavior can be also noticed in the load-displacement evolution curve, Figure 15c, which in comparison with the previous case, the failure initiation is predicted to occur at around $u_y/\bar{u}_y = 0.5$.

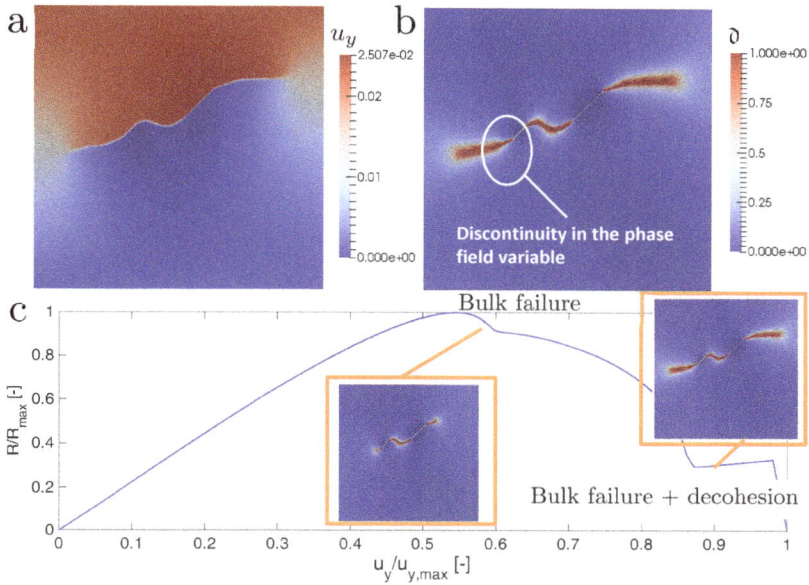

Figure 14. Specimen with 2 inclined flaws and a horizontal joint. Case $\bar{u}_y = 0.5\bar{u}_x$ traction-traction: (**a**) vertical displacement; (**b**) damage pattern; (**c**) Load-displacement evolution curve including snapshots of the phase field variable.

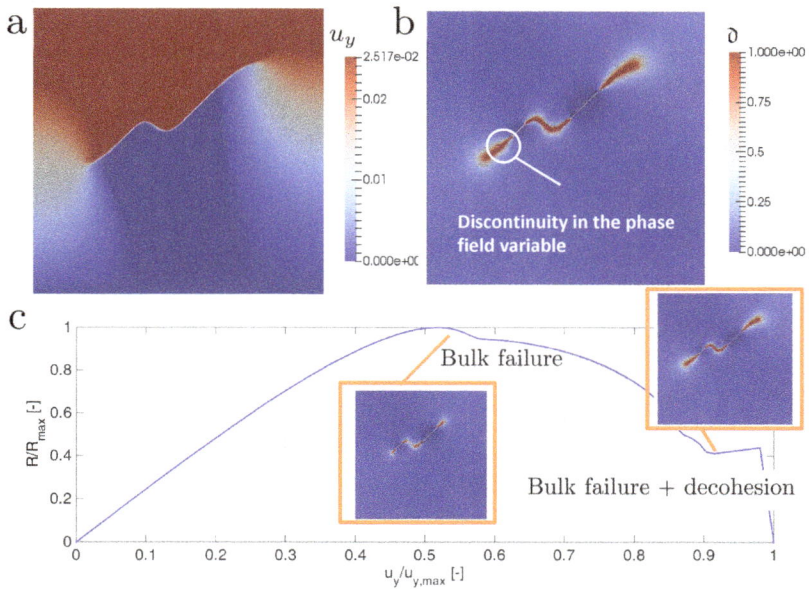

Figure 15. Specimen with 2 inclined flaws and a horizontal joint. Case $\bar{u}_y = \bar{u}_x$ traction-traction: (**a**) vertical displacement; (**b**) damage pattern; (**c**) Load-displacement evolution curve including snapshots of the phase field variable.

5.3.2. Specimen with Three Inclined Flaws

The previous analysis can be further extended to specimen definitions with 3 parallel flaws, 2 in the topmost substrate and 1 in the bottommost substrate. Moreover, regarding the joint position, we analyze the cases given by: (1) a single horizontal joint whose position is coincident with that described in the previous sections (Figure 16), and (2) a single vertical joint, placed at a distance of 20 mm with respect to the rightmost vertical edge, see Figure 18a.

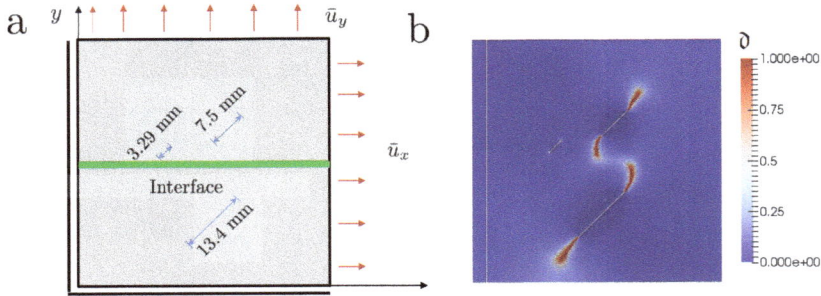

Figure 16. (**a**) Specimen with 3 inclined flaws and a horizontal joint. (**b**) Phase field damage pattern at the end of the simulation.

With reference to the case with the horizontal interface under traction-traction loading conditions given by $\bar{u}_y = \bar{u}_x$, it can be observed that the inclusion of the third flaw in the bottommost substrate of the model notably modifies the damage pattern. It is also noticed that the smallest flaw is not activated by the applied loading, and no damage events occur at the corresponding tips (Figure 16). Thus, crack phenomena are associated with the remaining larger flaws, featuring clear propagation paths which tend to coalesce. However, such two main cracks interact with the existing interface, and correspondingly the final damage pattern is a combination of bulk and interface failure. This sequence of damage can be identified analyzing the load-displacement curve (Figure 17) where different snapshots of the PF variable and the horizontal displacement are given. According to this graph, as usual, an initial linear evolution is depicted up to the fracture propagation of the main two flaws, which coincide with the first peak of the curve. Current results predict that these two main cracks initially propagate without involving any decohesion along the interface (note the continuity of the horizontal displacement field). However, when these two cracks are very close to the interface, they also induce the development of decohesion events, which are responsible for the final rupture of the specimen. This fact is characterized by the final abrupt failure in the load-displacement curve due to the tension cut-off interface model herein envisaged.

Regarding the vertical joint configuration, Figure 18b,c depict the final PF patterns for the cases $\bar{u}_y = -0.1\bar{u}_x$ and $\bar{u}_y = -\bar{u}_x$, respectively, showing completely different failure mechanisms in the predictions. Clear bulk failure emanating from the two larger flaws can be observed for the case $\bar{u}_y = -\bar{u}_x$, whereas the scenario setting $\bar{u}_y = -0.1\bar{u}_x$ shows that the main failure mechanism in this later specimen is due to joint decohesion. These two failure patterns can be confirmed by the analysis of the corresponding load-displacement curves, see Figure 19. Thus, the case $\bar{u}_y = -0.1\bar{u}_x$ features a standard evolution up to the first peak. At this point, the decohesion failure starts in unstable manner, followed by a second linear evolution up to the complete decohesion without any bulk failure, see the snapshot of the vertical displacements in Figure 19b. This response contrasts with that corresponding to the case $\bar{u}_y = -\bar{u}_x$, whereby the failure mechanism is due to the bulk failure without any interaction with the interface decohesion.

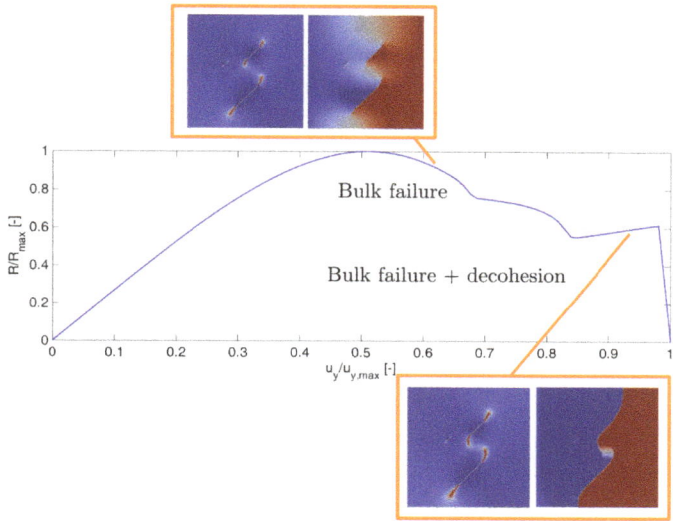

Figure 17. Specimen with 3 inclined flaws and a horizontal joint. Load-displacement evolution curve including snapshots of the bulk (phase field) failure and the horizontal displacement fields.

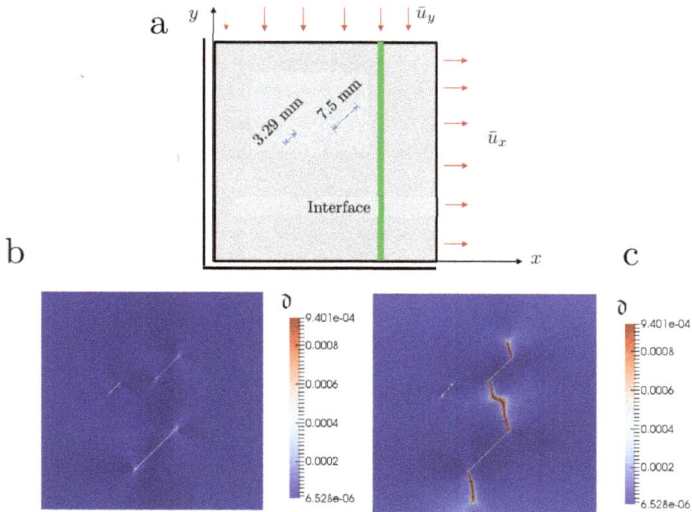

Figure 18. Specimen with 3 inclined flaws and a vertical interface: (**a**) Loading and geometric description. (**b**) Phase field pattern for $\bar{u}_y = -0.1\bar{u}_x$. (**c**) Phase field pattern for $\bar{u}_y = -\bar{u}_x$.

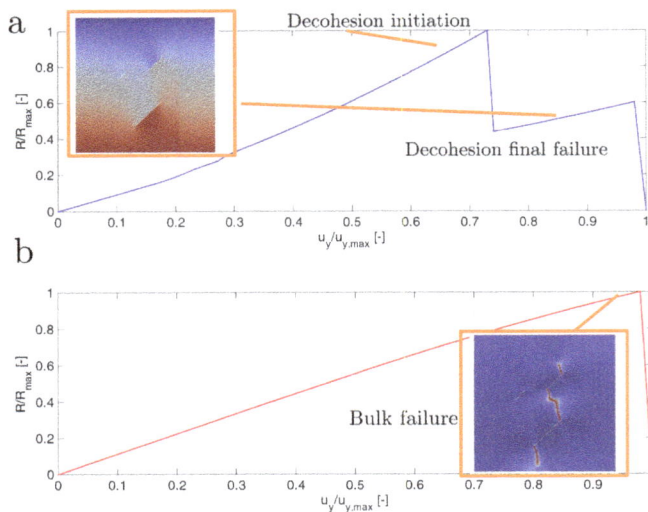

Figure 19. Specimen with 3 inclined flaws and a vertical interface: (**a**) Load-displacement curve for $\bar{u}_y = -0.1\bar{u}_x$. (**b**) Load-displacement curve for $\bar{u}_y = -\bar{u}_x$.

6. Conclusions

In this paper, the application of the combined PF-CZ approach for fracture heterogeneous in rock masses has been outlined with the incorporation of geo-mechanical parameters into the interface definition. The present methodology has been first validated for matrix rock damage, exhibiting an excellent accuracy with respect to experimental data. Subsequently, the developed method has been applied to the simulation of complex damage events in heterogeneous rock masses with weak joints and multiple flaws.

Results from the current model illustrate that the PF-CZ approach permits the prediction of crack initiation, propagation, coalescence, and branching in geo-materials in an autonomous manner without the necessity of user intervention. Moreover, differing from alternative PF formulations, its coupling with interface-like modeling techniques allows the characterization of tortuous crack patterns along predefined weak joints in such materials.

Future directions in this concern will regard the extension of the proposed method for anisotropic rocks, hydraulic fracture, porous media, and modified formulations for triggering shear failure.

Author Contributions: Conceptualization, J.R., P.D.; Methodology, J.R., M.P.; Investigation, P.R. Budarapu, J.R.; Resources, J.R.; Writing-Original Draft Preparation, J.R., P.R.B., M.P.; Writing-Review and Editing, J.R., P.R.B., M.P., P.D.; Funding Acquisition, J.R.

Funding: J. Reinoso is grateful to the support of the Spanish Ministry of Economy and Competitiveness/FEDER (Project MAT2015-71036-P).

Conflicts of Interest: The authors declare no conflict of interest.

References

1. Hoek, E.; Bieniawski, Z.T. Brittle Fracture Propagation in Rock under Compression. *Int. J. Fract.* **1984**, *26*, 276–294. [CrossRef]
2. Hoek, E. Brittle failure of rocks in rock mechanics in engineering practice. In *Rock Mechanics in Engineering Practice*; Stagg, K.G., Zienkiewicz, O.C., Eds.; Wiley: New York, NY, USA, 1968; pp. 99–124.
3. Lajtai, E.Z. Brittle Fracture in Compression. *Int. J. Fract.* **1974**, *10*, 525–536. [CrossRef]

4. Ingraffea, A.R.; Heuze, F.E. An Analysis of Discrete Fracture Propagation in Rock Loaded in Compression. In Proceedings of the 18th U.S. Symposium on Rock Mechanics, Keystone, Colorado, CO, USA, 22–24 June 1977.

5. Suits, L.D.; Sheahan, T.C.; Wong, L.N.Y.; Einstein, H.H. Using high speed video imaging in the study of cracking processes in rock. *Geotech. Test. J.* **2009**, *32*, 164–180. [CrossRef]

6. Bobet, A. Numerical simulation of initiation of tensile and shear cracks. In Proceedings of the DC Rocks (2001) the 38th US Symposium on Rock Mechanics (USRMS), Washington, DC, USA, 7–10 July 2011.

7. Yang, S.Q.; Yang, D.S.; Jing, H.W.; Li, Y.H.; Wang, S.Y. An experimental study of the fracture coalescence behaviour of brittle sandstone specimens containing three fissures. *Rock Mech. Rock Eng.* **2012**, *45*, 563–582. [CrossRef]

8. Reyes, O.; Einstein, H.H. Failure mechanisms of fractured rock—A fracture coalescence model. In Proceedings of the 7th International Conference on Rock Mechanics, International Society for Rock Mechanics, Aachen, Germany, 16–20 September 1991.

9. Wong, L.; Einstein, H. Crack coalescence in molded gypsum and carrara marble: Part 1. Macroscopic observations and interpretation. *Rock Mech. Rock Eng.* **2009**, *42*, 475–511. [CrossRef]

10. Wong, R.; Chau, K.; Tang, C.; Lin, P. Analysis of crack coalescence in rock-like materials containing three flaws—Part I: Experimental approach. *Int. J. Rock Mech.* **2001**, *38*, 909–924. [CrossRef]

11. Sagong, M.; Bobet, A. Coalescence of multiple flaws in a rock-model material in uniaxial compression. *Int. J. Rock Mech. Min. Sci.* **2002**, *39*, 229–241. [CrossRef]

12. Ma, G.W.; Dong, Q.Q.; Fan, L.F.; .Gao, J. An investigation of non-straight fissures cracking under uniaxial compression. *Eng. Fracture Mechanics* **2018**, *15*, 300–310. [CrossRef]

13. Rabczuk, T.; Belytschko, T. Cracking particles: A simplified meshfree method for arbitrary evolving cracks. *Int. J. Numer. Meth. Eng.* **2004**, *61*, 2316–2343. [CrossRef]

14. Rabczuk, T.; Belytschko, T. A three-dimensional large deformation meshfree method for arbitrary evolving cracks. *Comput. Methods Appl. Mech. Eng.* **2007**, *196*, 2777–2799. [CrossRef]

15. Rabczuk, T.; Zi, G.; Bordas, S.; Nguyen-Xuan, H. A simple and robust three-dimensional cracking-particle method without enrichment. *Comput. Methods Appl. Mech.* **2010**, *199*, 2437–2455. [CrossRef]

16. Rabczuk, T.; Ren, H. A peridynamics formulation for quasi-static fracture and contact in rock. *Eng. Geol.* **2017**, *225*, 42–48. [CrossRef]

17. Ren, H.; Zhuang, X.; Cai, Y.; Rabczuk, T. Dual-horizon peridynamics. *Int. J. Numer. Meth. Eng.* **2016**, *108*, 1451–1476. [CrossRef]

18. Ren, H.; Zhuang, X.; Rabczuk, T. Dual-horizon peridynamics: A stable solution to varying horizons. *Comput. Methods Appl. Mech. Eng.* **2017**, *318*, 762–782. [CrossRef]

19. Bobet, A.; Einstein, H.H. Fracture coalescence in rock-type materials under uniaxial and biaxial compression. *Int. J. Rock Mech. Min. Sci.* **1998**, *35*, 863–888. [CrossRef]

20. Gonçalves da Silva, B.; Einstein, H.H. Modeling of crack initiation, propagation and coalescence in rocks. *Int. J. Fract.* **2013**, *182*, 167–186. [CrossRef]

21. Lemaitre, J.; Desmorat, R. *Engineering Damage Mechanics: Ductile, Creep, Fatigue and Brittle Failures*; Springer-Verlag: Berlin, Germany, 2005.

22. Bažant, Z.P.; Pijaudier-Cabot, T.G.P. Nonlocal continuum damage, localization instability and convergence. *J. Appl. Mech.* **1988**, *55*, 287–293. [CrossRef]

23. Jirásek, M. Nonlocal models for damage and fracture: Comparison of approaches. *Int. J. Solids Struct.* **1988**, *35*, 4133–4145. [CrossRef]

24. Forest, S. Micromorphic approach for gradient elasticity, viscoplasticity, and damage. *J. Engnr. Mech.* **2009**, *35*, 117–131. [CrossRef]

25. Dimitrijevic, B.J.; Hackl, K. A regularization framework for damage-plasticity models via gradient enhancement of the free energy. *Int. J. Numer. Methods Biomed. Energy* **2011**, *27*, 1199–1210. [CrossRef]

26. Peerlings, R.; Geers, M.; de Borst, R.; Brekelmans, W. A critical comparison of non local and gradient-enhanced softening continua. *Int. J. Solids Struct.* **2001**, *38*, 7723–7746. [CrossRef]

27. Areias, P.; Msekh, M.A.; Rabczuk, T. Damage and fracture algorithm using the screened Poisson equation and local remeshing. *Eng. Fract. Mech.* **2016**, *158*, 116–143. [CrossRef]

28. Moës, N.; Dolbow, J.; Belytschko, T. A finite element method for crack growth without remeshing. *Int. J. Numer. Methods Eng.* **1999**, *46*, 131–150. [CrossRef]

29. Dolbow, J.; Moeës, N.; Belytschko, T. An extended finite element method for modeling crack growth with contact. *Comput. Meth. Appl. Mech. Eng.* **2001**, *190*, 6825–6846. [CrossRef]

30. Fries, T.P.; Belytschko, T. The extended/generalized finite element method: An overview of the method and its applications. *Int. J. Numer. Methods Eng.* **2010**, *84*, 253–304. [CrossRef]

31. Simo, J.C.; Oliver, J.; Armero, F. An analysis of strong discontinuities induced by strainsoftening in rate-independent inelastic solids. *Comput. Mech.* **1993**, *12*, 277–296. [CrossRef]

32. Linder, C.; Armero, F. Finite elements with embedded strong discontinuities for the modeling of failure in solids. *Int. J. Numer. Methods Eng.* **2007**, *72*, 1391–1433. [CrossRef]

33. Oliver, J.; Huespe, A.; Blanco, S.; Linero, D. Stability and robustness issues in numerical modeling of material failure with the strong discontinuity approach. *Comput. Methods Appl. Mech. Eng.* **2006**, *195*, 7093–7114. [CrossRef]

34. Armero, F.; Linder, C. New finite elements with embedded strong discontinuities for finite deformations. *Comput. Methods Appl. Mech. Eng.* **2008**, *197*, 3138–3170. [CrossRef]

35. Xu, X.P.; Needleman, A. Numerical simulation of fast crack growth in brittle solids. *J. Mech. Phys. Solids* **1994**, *42*, 1397–1434. [CrossRef]

36. Ortiz, M.; Pandolfi, A. Finite deformation irreversible cohesive elements for three-dimensional crack-propagation analysis. *Int. J. Num. Meth. Eng.* **1999**, *44*, 1267–1282. [CrossRef]

37. Park, K.; Paulino, G.H.; Roesler, J.R. A unified potential-based cohesive model of mixed-mode fracture. *J. Mech. Phys. Solids* **2009**, *57*, 891–908. [CrossRef]

38. Paggi, M.; Wriggers, P. Stiffness and strength of hierarchical polycrystalline materials with imperfect interfaces. *J. Mech. Phys. Solids* **2012**, *60*, 557–572. [CrossRef]

39. Reinoso, J.; Paggi, M. A consistent interface element formulation for geometrical and material nonlinearities. *Comput. Mech.* **2014**, *54*, 1569–1581. [CrossRef]

40. Infuso, A.; Corrado, M.; Paggi, M. Image analysis of polycrystalline solar cells and modelling of intergranular and transgranular cracking. *J. Eur. Ceram. Soc.* **2015**, *34*, 2713–2722. [CrossRef]

41. Paggi, M.; Reinoso, J. An anisotropic large displacement cohesive zone model for fibrillar and crazing of interfaces. *Int. J. Solids Struct.* **2015**, *69*, 106–120. [CrossRef]

42. Francfort, G.A.; Marigo, J.J. Revisiting brittle fracture as an energy minimization problem. *J. Mech. Phys. Solids* **1998**, *46*, 1319–1342. [CrossRef]

43. Griffith, A.A. The phenomena of rupture and flow in solids. *Philos. Trans. Royal Soc. Lond. A* **1921**, *221*, 163–198. [CrossRef]

44. Ambrosio, L.; Tortorelli, V.M. Approximation of functional depending on jumps by elliptic functional via C-convergence. *Commun. Pure Appl. Math.* **1990**, *43*, 999–1036. [CrossRef]

45. Ambrosio, L.; Tortorelli, V.M. On the approximation of free discontinuity problems. *Boll. Union Mat. Ital.* **1992**, *6*, 105–123.

46. Dal Maso, G. Introduction to Γ-Convergence. Progress in Nonlinear Differential Equations and Their Applications, Birkhäuser. Progress in Nonlinear Differential Equations and Their Applications. Available online: https://link.springer.com/book/10.1007/978-1-4612-0327-8 (accessed on 13 March 2019).

47. Bourdin, B.; Francfort, G.A.; Marigo, J.J. Numerical experiments in revisited brittle fracture. *J. Mech. Phys. Solids* **2000**, *48*, 797–826. [CrossRef]

48. Bourdin, B.; Francfort, G.A.; Marigo, J.J. The variational approach to fracture. *J. Elast.* **2008**, *91*, 145–148. [CrossRef]

49. Miehe, C.; Hofacker, M.; Welschinger, F. A phase field model for rateindependent crack propagation: Robust algorithmic implementation based on operator splits. *Comput. Methods Appl. Mech. Eng.* **2010**, *199*, 2765–2778. [CrossRef]

50. Miehe, C.; Welschinger, F.; Hofacker, M. Thermodynamically consistent phasefield models of fracture: Variational principles and multi-field fe-implementations. *Int. J. Numer. Methods Eng.* **2010**, *83*, 1273–1311. [CrossRef]

51. Kuhn, C.; Müller, R. A continuum phase field model for fracture. *Eng. Fract. Mech.* **2010**, *77*, 3625–3634. [CrossRef]

52. Borden, M.J.; Hughes, T.J.R.; Landis, C.M.; Verhoosel, C.V. A higher-order phase-field model for brittle fracture: Formulation and analysis within the isogeometric analysis framework. *Comput. Methods Appl. Mech. Eng.* **2014**, *273*, 100–118. [CrossRef]

53. Amiri, F.; Millán, D.; Shen, Y.; Rabczuk, T.; Arroyo, M. Phase-field modeling of fracture in linear thin shells. *Theor. Appl. Fract. Mech.* **2014**, *69*, 102–109. [CrossRef]
54. Areias, P.; Rabczuk, T.; Msekh, M.A. Phase-field analysis of finite-strain plates and shells including element subdivision. *Comput. Methods Appl. Mech. Eng.* **2016**, *312*, 322–350. [CrossRef]
55. Reinoso, J.; Paggi, M.; Linder, C. Phase field modeling of brittle fracture for enhanced assumed strain shells at large deformations: Formulation and finite element implementation. *Comp. Mech.* **2017**. [CrossRef]
56. Ulmer, H.; Hofacker, M.; Miehe, C. Phase field modeling of brittle and ductile fracture. *Proc. Appl. Math. Mech.* **2013**, *13*, 533–536. [CrossRef]
57. Ambati, M.; Gerasimov, T.; de Lorenzis, L. Phase-field modeling of ductile fracture. *Comp. Mech.* **2015**, *55*, 1–24. [CrossRef]
58. Verhoosel, C.; de Borst, R. A phase-field model for cohesive fracture. *Int. J. Numer. Meth. Eng.* **2013**, *96*, 43–62. [CrossRef]
59. Borden, M.J.; Verhoosel, C.V.; Scott, M.A.; Hughes, T.J.R.; Landis, C.M. A phase-field description of dynamic brittle fracture. *Comput. Methods Appl. Mech. Eng.* **2012**, *217–220*, 77–95. [CrossRef]
60. Hofacker, M.; Miehe, C. A phase field model of dynamic fracture: Robust field updates for the analysis of complex crack patterns. *Int. J. Numer. Methods Eng.* **2013**, *93*, 276–301. [CrossRef]
61. Miehe, C.; Schanzel, L.; Ulmer, H. Phase field modeling of fracture in multi-physics problems. Part I. Balance of crack surface and failure criteria for brittle crack propagation in thermo-elastic solids. *Comput. Methods Appl. Mech. Eng.* **2015**, *294*, 449–485. [CrossRef]
62. Miehe, C.; Schanzel, L.; Ulmer, H. Phase field modeling of fracture in multi-physics problems. Part II. Coupled brittle-to-ductile failure criteria and crack propagation in thermo-elastic solids. *Comput. Methods Appl. Mech. Eng.* **2015**, *294*, 486–522. [CrossRef]
63. Martínez-Pañeda, E.; Golahmar, A.; Niordson, C.F. A phase field formulation for hydrogen assisted cracking. *Comput. Methods Appl. Mech. Eng.* **2018**, *342*, 742–761. [CrossRef]
64. Zhou, S.; Zhuang, X.; Zhu, H.; Rabczuk, T. Phase field modelling of crack propagation, branching and coalescence in rocks. *Theor. Appl. Fract. Mech.* **2018**, *96*, 174–192. [CrossRef]
65. Zhang, X.; Sloan, S.W.; Vignes, C.; Sheng, D. A modification of the phase-field model for mixed mode crack propagation in rock-like materials. *Comput. Methods Appl. Mech. Eng.* **2017**, *322*, 123–136. [CrossRef]
66. Bryant, E.C.; Sun, W. A mixed-mode phase field fracture model in anisotropic rocks with consistent kinematics. *Comput. Methods Appl. Mech. Eng.* **2018**, *342*, 561–584. [CrossRef]
67. Li, L.; Xia, Y.; Huang, B.; Zhang, L.; Li, M.; Li, A. The Behaviour of Fracture Growth in Sedimentary Rocks: A Numerical Study Based on Hydraulic Fracturing Processes. *Energies* **2016**, *9*, 169. [CrossRef]
68. Paggi, M.; Reinoso, J. Revisiting the problem of a crack impinging on an interface: A modeling framework for the interaction between the phase field approach for brittle fracture and the interface cohesive zone model. *Comput. Methods Appl. Mech. Eng.* **2017**, *321*, 145–172. [CrossRef]
69. Carpinteri, A.; Paggi, M. Size-scale effects on strength, friction and fracture energy of faults: A unified interpretation according to fractal geometry. *Rock Mech. Rock Eng.* **2008**, *41*, 735–746. [CrossRef]
70. Wu, J.Y. A unified phase-field theory for the mechanics of damage and quasi-brittle failure in solids. *J. Mech. Phys. Solids* **2017**, *103*, 72–99. [CrossRef]
71. Wu, J.Y.; Nguyen, V.P. A length scale insensitive phase-field damage model for brittle fracture. *J. Mech. Phys. Solids* **2018**, *119*, 20–42. [CrossRef]
72. Nguyen, V.P.; Wu, J.Y. Modeling dynamic fracture of solids using a phase-field regularized cohesive zone model. *Comput. Methods Appl. Mech. Eng.* **2018**, *340*, 1000–1022. [CrossRef]
73. Amor, H.; Marigo, J.J.; Maurini, C. Regularized formulation of the variational brittle fracture with unilateral contact: Numerical experiments. *J. Mech. Phys. Solids* **2009**, *57*, 1209–1229. [CrossRef]
74. Msekh, M.A.; Sargado, M.; Jamshidian, M.; Areias, P.; Rabczuk, T. Abaqus implementation of phase-field model for brittle fracture. *Comput. Mater. Sci.* **2015**, *96*, 472–484. [CrossRef]
75. Lei, Q.; Latham, J.P.; Xiang, J. Implementation of an emprirical joint constitutive model into finite-discrete element analysis of the geomechanical behavior of fractured rocks. *Rock Mech. Rock Eng.* **2016**, *49*, 4799–4816. [CrossRef]
76. Barton, N.; Choubey, V. The shear strength of rock joints in theory and practice. *Rock Mech.* **1977**, *10*, 1–54. [CrossRef]

77. Barton, N. Shear strength criteria for rock, rock joints and rock masses: Problems and some solutions. *J. Rock Mech. Geotech. Eng.* **2013**, *5*, 249–261. [CrossRef]
78. Helgeson, D.E.; Aydin, A. Characteristics of joint propagation across layer interfaces in sedimentary rocks. *J. Struct. Geol.* **1991**, *13*, 897–911. [CrossRef]
79. Li, Y.; Liu, W.; Yang, C.; Daemen, J.J.K. Experimental investigation of mechanical behavior of bedded rock salt containing inclined interlayer. *Int. J. Rock Mech. Min. Sci.* **2014**, *69*, 39–49. [CrossRef]
80. Tromans, D.; Meech, J.A. Fracture toughness and surface energies of minerals: Theoretical estimates for oxides, sulphides, silicates and halides. *Miner. Eng.* **2002**, *15*, 1027–1041. [CrossRef]

MDPI

St. Alban-Anlage 66

4052 Basel

Switzerland

Tel. +41 61 683 77 34

Fax +41 61 302 89 18

www.mdpi.com

Energies Editorial Office

E-mail: energies@mdpi.com

www.mdpi.com/journal/energies

www.ingramcontent.com/pod-product-compliance
Lightning Source LLC
Chambersburg PA
CBHW051912210326
41597CB00033B/6117